T0335772

Springer Environmental Science and Engineering

More information about this series at http://www.springer.com/series/10177

Dahe Qin · Yongjian Ding · Mu Mu
Editors

Climate and Environmental Change in China: 1951–2012

Editors
Dahe Qin
Cold and Arid Regions Environmental and Engineering Research Institute
Chinese Academy of Sciences
Lanzhou
China

and

China Meteorological Administration
Beijing
China

Yongjian Ding
Cold and Arid Regions Environmental and Engineering Research Institute
Chinese Academy of Sciences
Lanzhou
China

Mu Mu
Institute of Oceanology
Chinese Academy of Sciences
Qingdao
China

ISSN 2194-3214 ISSN 2194-3222 (electronic)
Springer Environmental Science and Engineering
ISBN 978-3-662-48480-7 ISBN 978-3-662-48482-1 (eBook)
DOI 10.1007/978-3-662-48482-1

Library of Congress Control Number: 2015950033

Springer Heidelberg New York Dordrecht London

Printed on acid-free paper

Springer-Verlag GmbH Berlin Heidelberg is part of Springer Science+Business Media
(www.springer.com)

Preface

Global climate and environmental change is one of the key scientific and societal issues worldwide. The Intergovernmental Panel on Climate Change (IPCC) has published five assessment reports since 1990. The first scientific assessment report of "China Climate and Environmental Change" published in 2005. Subsequently, the first "National Assessment Report on Climate Change" in China was published in 2007 based on the results from the scientific assessment report. These two assessment reports provided important scientific basis for full recognition of China's climate and environmental change, impacts, and adaptation, attracting much attention from scientific community, governmental agencies, policy makers, and general public as a whole.

In order to further coordinate with the international climate change assessment, synthesize the up-to-date results of climate and environmental change research, and promote the Chinese scientists' contribution to global change science, China launched the second assessment of "Climate and Environmental Change in China: 2012," supported by the Chinese Academy of Sciences and the China Meteorological Administration in 2008. The original plan was that the assessment in English should be completed and published in 2012, in line with the completion and publication of the IPCC Fifth Assessment Report in 2013. The Chinese version of "Climate and Environmental Change in China: 2012" was indeed published in 2012 with four volumes: the scientific basis; the impact of vulnerability; mitigation and adaptation of climate and environmental change in China; and a technical summary volume. This English version is a synthesis of the "China climate and environmental Change: 2012" with some additional background of climate, environmental, and geographic information.

This book adopts the procedures and styles of international scientific assessment on climate and environmental change. Based on the most up-to-date scientific research results from national and international literatures, the authors extracted the mainstream views and summarized the major conclusions with concessions from scientific community. All cited references are peer-reviewed articles, especially those articles published in Chinese journals which have limited availability to the international scientific community. Based on community-recognized major datasets on conclusions, this book provides a comprehensive analysis and assessment on climate and the environment changes, and its impacts, adaptation, and mitigation measures. The aim of the "Climate and Environmental Change in China: 1951–2012" is to introduce the most up-to-date climate change research results by Chinese scientists to international community, enhancing a better understanding of progress on developments of climate change science in China and jointly dealing with the impacts of climate change worldwide.

This book consists of eight chapters. Chapter 1 provides background information on the basic characteristics of climate and environment conditions, and societal and economic developments in China. Chapter 2 introduces the observed evidence of climate and environmental change during the past six decades, followed by discussions on the causes and uncertainties of climate and environmental change in Chap. 3 and on the impacts of climate change on China's environment, economy, and society in Chap. 4. Chapter 5 describes the projection of climate change by the end of twenty-first century and further discusses the possible implications. Chapter 6 briefly introduces the technologies and policy options on adaptation and mitigation of the impacts of climate and environmental change. Chap. 7 gives the Chinese government's strategic plans to address climate change. Finally, Chap. 8 summarizes the key findings and important scientific understanding.

The above contents include many aspects of sciences in climate and environmental change, such as in natural, societal, economic, and human dimensions. It is the most updated and authoritative scientific assessment report on understanding climate and environmental change, impacts, adaptation, and mitigation pathways in China. I am so pleased and excited to the publication of this book in English to the world climate research community and general public at large.

There are many colleagues and experts who contributed to this book. They are from the Chinese Academy of Sciences, the China Meteorological Administration, the Ministry of Education, the Ministry of Health, the Ministry of Water Resources, the Oceanic Administration, the Ministry of Agriculture and Forestry, the Development and Reform Commission, the Ministry of Foreign Affairs, the Ministry of Finance, the Chinese society Academy of Sciences, and societal organizations. Publication of this book is a truly team effort with experts from multiple organizations and disciplines. It is a contribution by Chinese climate scientists to better understand world climate and environmental changes.

I would like to express my gratitude to all authors and contributors, reviewer experts, project office, and secretarial group for their hard work and for taking responsibilities. The State Key Laboratory of Cryosphere Science is in charge of the project office and secretary group work. The secretary group consists of Wenhua Wang, Yawei Wang, Aihong Xie, Chuancheng Zhao, Jianbin Xiong, and Fu Sha, and they all have done a lot of fruitful work. I thank Lisa Fan who has done careful work that assured the book quality.

The financial support is in part provided by Ministry of Science and Technology of the People's Republic of China (Project NO.: 2013CBA01800-A), the Chinese Academy of Sciences (Project No.: KZCX2-XB2-16), and the China

Meteorological Administration. My gratitude also goes to Dr. Chunli Bai, the president of the Chinese Academy of Sciences, and Dr. Guoguang Zheng, the head of China Meteorological Administration for their understanding and support for this work.

Due to the complexity and uncertainty of climate and environment research, errors and mistakes may have occurred as well as imperfections in the layout of the book. The editors and chapter authors look forward to comments and suggestions from the readers. Criticisms from readers are the power to improve scientific research and assessment work in future.

Beijing
October 2014

Dahe Qin

Contents

Chapter 1
Climate, Environmental, and Socioeconomic Characteristics of China

Dahe Qin, Shu Tao, Suocheng Dong and Yong Luo

Abstract The features of the natural environment in China were described in this chapter, including topography, hydrology and water resources, cryospheric resources, plants, animals and biodiversity, soil types and land resources, and ocean and islands. After that, an eco-geographical region system of depicting zonal distribution of physiographical factors was introduced. China's climate is governed by its geography and topography, with most of the territory falling in the monsoon zone. The most direct component influencing China's climate is the East Asian atmospheric circulation (monsoon). Climate zoning in China provides a service for targeted industrial and agricultural activities, and socioeconomic development, and also a scientific basis for climate change adaptation. However, China is susceptible

The following researchers have also contributed to this chapter

Fei Li

Institute of Geographic Sciences and Natural Resources Research, Chinese Academy of Sciences, Beijing 100101, China

Wangzhou Yang

College of Economics and Management, Yunnan Normal University, Kunming 650500, China

D. Qin
State Key Laboratory of Cryospheric Sciences, Cold and Arid Regions Environmental and Engineering Research Institute, Chinese Academy of Sciences, Lanzhou 730000, China
e-mail: qdh@cma.gov.cn

D. Qin
China Meteorological Administration, Beijing 100081, China

S. Tao
College of Urban and Environmental Sciences, Peking University, Beijing 100871, China

S. Dong (✉)
Institute of Geographic Sciences and Natural Resources Research, Chinese Academy of Sciences, Beijing 100101, China
e-mail: dongsc@igsnrr.ac.cn

Y. Luo
Center for Earth System Science, Tsinghua University, Beijing 100084, China
e-mail: Yongluo@mail.tsinghua.edu.cn

D. Qin et al. (eds.), *Climate and Environmental Change in China: 1951–2012*,
Springer Environmental Science and Engineering,
DOI 10.1007/978-3-662-48482-1_1

to weather and climatic disasters. As one of the world's emerging economies, China has experienced rapid and stable socioeconomic development, industrialization, urbanization, and further internationalization since the introduction of reform and opening-up policy, ranking second of gross domestic product (GDP) around the world. At present, China is confronted with two main challenges: addressing climate change in the international arena and protecting its resources and environment with a domestic socioeconomic transition.

Keywords Physiographical factors · Regionalization · China climate · Climate resources · Climate zoning · Weather and climatic disasters · Socioeconomy · Population · Urbanization

1.1 Features of the Natural Environment in China

1.1.1 Geographical Location and Landmass of China

China is located on the eastern side of the Eurasian continent and on the west shore of the Pacific Ocean. From north to south, China spans ~5500 km, stretching from the center of the Heilongjiang River north of the town of Mohe in Heilongjiang Province (north of 53ºN) to the southernmost Zengmu Reef (near 4ºN); from east to west, the nation extends from the confluence of the Heilongjiang River and Wusuli River (west of 135ºE) to the Pamir Plateau in the Xinjiang Uygur Autonomous Region (near 73ºE), spanning ~5000 km. China has a land area of ~9.6 million km^2 and is the third largest country in the world, after Russia and Canada. China borders on 14 countries and shares a maritime border with eight countries. China is bordered by North Korea and Russia to the northeast; Kazakhstan, Kyrgyzstan, and Tadzhikistan to the northwest; Mongolia to the north; Afghanistan and Pakistan to the west; India, Nepal, and Bhutan to the southwest; and Myanmar, Laos, and Vietnam to the south.

1.1.2 Topography of China

China has a "terraced" terrain that gradually descends from the west to the east stepwise in three terraces, linking the mainland and the Pacific Ocean basin through a wide continental shelf. The highest terrace is the Qinghai-Tibet Plateau (the highest plateau in the world), with an average elevation of more than 4000 m and an area of ~250 million km^2. The Qinghai-Tibet Plateau is composed of seven mountain ranges, from north to south: the Kunlun Mountains, the Altun Mountains, the Qilian Mountains, the Tanggula Mountains, the Karakoram Mountains, the Gangdise Mountains, and the Himalaya Mountains. East of the Kunlun Mountains,

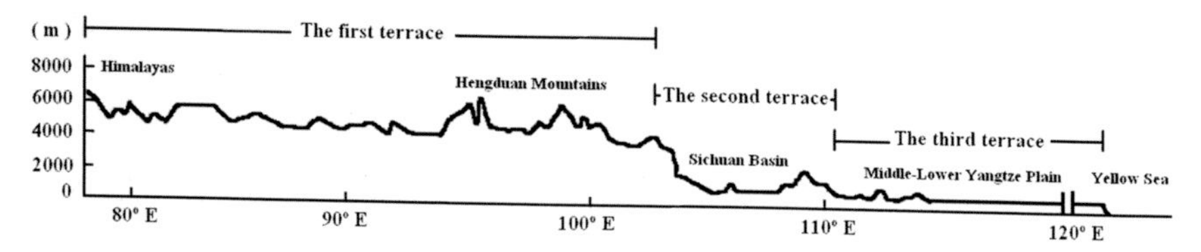

Fig. 1.1 Three terrain terraces in China

the second terrace includes the Qilian Mountains, the eastern rim of the Minshan Mountains, the Qionglai Mountains, and the Hengduan Mountains, and descends to an elevation of 1000–2000 m (some areas are below 500 m). The eastern rim of this second terrace is roughly bounded by the Greater Hinggan Mountains, the Taihang Mountains, the Wushan Mountains, the Wuling Mountains, and the Xuefeng Mountains. The second terrace consists of a series of mountains, plateaus, and basins including the Altai Mountains, the Tianshan Mountains, and the Qinling Mountains; the Inner Mongolia Plateau, the Loess Plateau, and the Yunnan-Guizhou Plateau; and the Junggar Basin, the Tarim Basin, the Qaidam Basin, and the Sichuan Basin, from north to south. The third terrace consists of hills and plains below 500 m in elevation and begins at a line from the Da Xing'an Ling Mountains to the Xuefeng Mountains and ranges eastward to the Pacific Ocean. Here, running from north to south are the Northeast Plain, the North China Plain, and the Middle-Lower Yangtze Plain; there is a wide, low, hilly region in the south Yangtze River area that is referred to as the Southeast Hills. Altitudes in the third terrace range from below 200 to 500 m, with only a few mountains reaching or exceeding 1000 m (Fig. 1.1).

Topography across China is varied and complicated, including plains, plateaus, mountains, hills, and basins. The proportion of various landforms is as follows: mountains, 33 %; plateaus, 26 %; basins, 19 %; plains, 12 %; and hills, 10 %. Including all mountainous areas and rugged plateaus, mountainous areas in China occupy fully two-thirds of total land area.

China has broad deserts and Gobi areas, mainly in the north, including arid and semiarid regions in the northwest and Inner Mongolia. Total desert areas comprise 1.28 million km^2, occupying 13 % of the country's area. The largest desert area is in the west of Wushaoling and Helan Mountains, and includes the Taklimakan Desert, the Gurbantunggut Desert, the Badan Jaran Desert, and the Tengger Desert (the four largest deserts in China). Around the periphery of large deserts, there is ribbon or ring of Gobi deserts.

1.1.3 Hydrology and Water Resources of China

There are more than 1500 rivers in China, each draining 1000 km^2 or more. Because most of the large rivers originate in the Qinghai-Tibet Plateau, China is rich in hydropower resources and leads the world in hydropower potential, with reserves of 680 million kW.

China's rivers can be categorized as exterior and interior systems. The catchment area for the exterior rivers, which empty into the ocean, accounts for 64 % of the country's total land area. The Yangtze, Yellow, Heilongjiang, Pearl, Liaohe, Haihe, Huaihe, and Lancang Rivers flow east, and empty into the Pacific Ocean. The Yarlungzangbo River in Tibet flows first east and then south into the Indian Ocean, and the Ertix River flows from the Xinjiang Uygur Autonomous Region to the Arctic Ocean. The catchment area for the interior rivers (those that flow into inland lakes or disappear into deserts or salt marshes) makes up 36 % of China's total land area. The Tarim River in southern Xinjiang is 2179 km long and is China's longest interior river.

China, however, is a country of a severe drought and water shortages. Total freshwater resources are 2.8 trillion m^3, accounting for 6 % of global water resources and ranking fourth in the world only after Brazil, Russia, and Canada. The total quantity of Chinese water resources is comparatively great, but its per capita average is only 2200 m^3, about 1/4 of the world average value and 1/5 of that of the USA. China's per capita water resources rank it in the 13 most water-poor countries in the world.

1.1.4 Cryosphere of China

The extensive cryospheric resources of China play an important role in the world. China has a large number of mountain glaciers in mid-latitude areas: 46,298 glaciers with a total area of 59,406 km^2, which is ~0.4 % of the total area of the world's glaciers and ice sheets, 14.6 % of the area of global mountain glaciers, and 47.6 % of the area of Asian mountain glaciers. The ice reserves of China are ~5590 km^3 (Table 1.1) (Liu et al. 2000).

Frozen ground in China can be divided into two categories: seasonally frozen ground and permafrost. Seasonally frozen ground occupies more than half of China's territory, with its southern boundary starting from Zhangfeng in Yunnan Province in the west, across Kunming and Guiyang, then around the northern margin of the Sichuan Basin, to Changsha, Anqing, and Hangzhou in the east. The depth of the seasonally frozen ground can be over 3 m in southern Heilongjiang Province, northeastern Inner Mongolia Autonomous Region, and northwestern Jilin Province, and the frozen depth gradually decreases as latitude decreases. The distribution areas of permafrost mainly include the Da Xing'an Ling Mountains and Xiao Xing'an Ling Mountains in the northeast, and the Altai Mountains, Tianshan Mountains, Qilian Mountains, and Tibetan Plateau in the west, accounting for 1/5 of China's land territory.

Frozen ground is divided into three regions in China: the East Region, the Northwest Region, and the Qinghai-Tibet Plateau Region. In the East Region, the distribution area is from the Da Xing'an Ling Mountains and Xiao Xing'an Ling Mountains in the north to the Yangtze River Basin and even to Zhejiang, Hunan, Fujian, and other provinces in some years. In the Northwest Region and the

Table 1.1 Numbers of glaciers in the mountain ranges of China (Liu et al. 2000)

Mountains	Elevation of peaks (m)	Number of glaciers		Area of glaciers		Ice reserves		Average area (km^2)	Coverage of glaciers (%)
		Number	%	km^2	%	km^3	%		
Altai	4374	403	0.87	280	0.47	16	0.28	0.68	0.97
Sawuershan	3835	21	0.05	17	0.03	1	0.01	0.8	0.38
Tianshan	7435	9081	19.61	9236	15.55	1012	18.1	1.02	4.36
Pamirs	7649	1289	2.78	2696	4.54	248	4.45	2.09	11.33
Karakoram[1]	8611	3454	7.46	6231	10.49	686	12.28	1.8	23.42
Kunlun	7167	7694	16.62	12266	20.65	1283	22.95	1.59	2.57
Altun	6295	235	0.51	275	0.46	16	0.28	1.17	0.49
Qilian	5827	2815	6.08	1931	3.25	93	1.67	0.69	1.46
Qiangtang Plateau	6822	958	2.07	1802	3.03	162	2.9	1.88	0.41
Tanggula	6621	1530	3.3	2213	3.73	184	3.29	1.45	1.57
Gangdise	7095	3538	7.64	1766	2.97	81	1.45	0.5	1.16
Nyenchen Tanglha	7162	7080	15.29	10701	18.01	1002	17.92	1.51	9.68
Hengduan	7514	1725	3.73	1580	2.66	97	1.74	0.92	0.44
Himalayas	8848	6475	13.99	8412	14.16	709	12.68	1.3	4.15
Total		46298	100	59406	100	5590	100	1.28	2.5

Note The 142 glaciers under the actual control of Palestine, located upstream of the Klechin River, which is the source of Yarkant River, are not included (these occupy an area of 609 km^2, and their ice reserves are 72 km^3)

Qinghai-Tibet Plateau Region, both seasonally frozen ground and permafrost are extensively distributed. Frozen ground is mainly distributed along latitudinal zones in the East Region and along altitudinal zones in the Qinghai-Tibet Plateau Region, with a combined distribution of the two types in the Northwest Region.

Permafrost in China can be divided into high-latitude permafrost and high-altitude permafrost. The former is located in the Northeast Region, while the latter is found in the mountains and plateaus in the west and also in high mountains in the east (such as the Huanggangliang Mountains of southern Da Xing'an Ling Mountains, the Changbai Mountains, the Wutai Mountains, and the Taibai Mountains). Seasonally frozen ground in China varies significantly within the year, showing distinct seasonal changes. The freeze begins in September and gradually expands from north to south. Both the area and the depth of the frozen ground in China reach a maximum in late winter and early spring, with more than 100 cm and even more than 200 cm in some regions of northern China and the Qinghai-Tibet Plateau. In summer, seasonally frozen ground shrinks continuously, declining to a minimum in August. Autumn and spring are transitional seasons. In autumn, the area and depth of frozen ground increase gradually, while these decrease gradually in the spring. Seasonally frozen area has the longest freezing period (over half a year) in the Da Xing'an Ling Mountains and Xiao Xing'an Ling Mountains and on the Qinghai-Tibet Plateau, while the shortest freezing period is in the Yangtze River Basin and the Huaihe River Basin, lasting about two or three months (Chen and Li 2008).

1.1.5 Animals, Plants, and Biological Diversity in China

China is rich in animal species, with 6222 kinds of vertebrates, including 2404 kinds of terrestrial vertebrates and 3862 species of fish, accounting for $\sim$10 % of vertebrate species in the world. China hosts hundreds of kinds of specific rare animals such as pandas, golden monkeys, the south China tiger, brown-eared pheasant, red-crowned crane, crested ibis, white-flag dolphin, and the Yangtze alligator.

China also has some of the richest plant resources in the world, with more than 32,000 kinds of vascular plants. China has more than 7000 kinds of woody plants, including more than 2800 kinds of trees, among which there are Chinese endemic tree species such as *Metasequoia glyptostroboides*, *Glyptostrobus pensilis*, silver fir, Chinese fir, golden larch, Taiwania, Fujian cypress, dove tree, *Eucommia ulmoides*, and *Camptotheca acuminate*. There are more than 2000 kinds of edible plants and more than 3000 kinds of medicinal plants found in China. Ginseng from the Changbai Mountains, safflower from Tibet, medlar from Ningxia, and *Panax notoginseng* from Yunnan and Guizhou are all famous medicinal plants.

China has a rich biodiversity, occupying a very unique position in the world. Biodiversity experts ranked China as 8th in the 12 most biodiverse countries of the world in 1990. China has the richest biodiversity among Northern Hemisphere countries.

1. High species richness: more than 30,000 species of vascular plants, third in the world after Brazil and Colombia.
2. Abundant endemic genera and species: China has 17,300 endemic species of vascular plants, accounting for more than 57 % of China's vascular plants. It has ~110 endemic mammalian species, occupying ~19 % of the 581 species in China.
3. Ancient flora: Many areas in China preserve ancient relict components of the Cretaceous and Tertiary (because China was not covered by glaciers in the Quaternary Ice Age). For example, China has six of the existing seven families of conifers in the world.
4. Abundant germplasm resources of cultivated plants, domesticated animals, and their wild relatives: China has thousands of years of cultivation history, and so it has the greatest abundance of cultivated plants and domesticated animals in the world.
5. Rich ecosystem types: China has various types of terrestrial ecosystems, including forest, scrub, steppe and savanna, meadow, desert, and alpine tundra. These ecosystems can be further divided into ~600 subecosystems.
6. Complicated and diverse spatial patterns: China has vast territory, undulating landforms, and complex and changeable climates. From north to south, climates range from a cold temperate zone, temperate zone, warm temperate zone, subtropical zone, and tropical zone. Chinese biomes consist of cold temperate coniferous forests, temperate coniferous and broad-leaved mixed forests, warm temperate deciduous broad-leaved forests, subtropical evergreen broad-leaved forests, and tropical seasonal rain forests. From east to west, along the track of reduced precipitation, the mixed coniferous broad-leaved forest and broad-leaved deciduous forest are sequentially replaced by the meadow steppe, typical steppe, desert steppe, steppe desert, typical desert, and dry desert in the north; in the south, there are obvious differences between the evergreen broad-leaved forests of the eastern subtropical zone in the Southern Hills and the evergreen broad-leaved forests of the western subtropical zone in the Yunnan-Guizhou Plateau.

1.1.6 Soil and Land Resources of China

1. Soil types and distribution

China's soil resources are rich, and its soil types are complex. Soil types in China can be classified into red soil, brown soil, cinnamon soil, black soil, chestnut soil, desert soil, fluvo-aquatic soil (including sand ginger chernozem), anthropogenic alluvial soil, paddy soil, wet soil (meadow, boggy soil), saline-alkali soil, endo-dynamorphic soil, and alpine soil. Among these soil series, the red soil series, which are the most important soil resource in tropical and subtropical regions, are suitable for cash crops, fruit trees, and lumber. From south to north, the spatial distribution of red soil series is latosol, dry red soil (savanna soil), lateritic red soil (laterization red soil), red soil, yellow soil, and other soil types.

The spatial distribution from south to north of brown soil series, which developed in humid forest areas of eastern China and are good for lumber production, is yellow brown soil, brown soil, dark brown soil, podzolic soil, and other soil types. The soil types in the cinnamon soil series, including cinnamon soil, dark loessial soil, and gray cinnamon soil, are mainly used as rain-fed land in northern China, except for gray cinnamon soil, which is used for forestland. The soil types in the black soil series, including gray black soil (gray forest soils), black soil, white pulp soil, and chernozem, are in temperate forests and grasslands, and suitable for the agriculture, animal husbandry, and forestry. The soil types in the chestnut soil series, including chestnut soil, brown calcic soil, and sierozem, which are a widely distributed grassland soil type in northern China, are used for animal husbandry.

The desert soil series consisted of gray desert soil, gray brown desert soil, brown desert soil, and takyr (important soil resources in the desert area of northwest China) are mainly used for animal husbandry and for agriculture in those areas with good water resources. The fluvo-aquatic soil types are mainly distributed in the Huang-Huai-Hai Plain, the Lower Liao River Plain, the Middle-Lower Yangtze Plain, and the Fen-Wei Valley, and are used in the cultivation of wheat, corn, sorghum, and cotton. Paddy soil types are mainly found south of the Qinling-Huai River and centered on the middle-lower Yangtze Plain, the Pearl River Delta, the Sichuan Basin, and the Taiwan Western Plain. Paddy soil is important in rice production.

2. Land Resources

China's land resources are vast, and its use types are diverse. Cultivated land, forests, and pasture account for a large portion of the useable land, with forests and grasslands being larger than the area of cultivated land. However, the current situation of China's land resources is deteriorating. Under the pressure of population growth and economic development, land resources are becoming increasingly smaller. This is exacerbated by extensive use of and serious waste of land resources, coupled with inefficient management.

Although the total land area of China is large, the per capita land area is small. China's total land area is $\sim 9.6 \times 10^6$ km^2. But its population accounts for 1/5 of the world's population, which makes its per capita land area far below the average for the rest of the world (it ranks 11 in the world's top 12 largest countries). Although China's land area is larger than Japan, Germany, and the UK, its unused land is vast. Much of the land cannot be used for agriculture or for construction. This further reduces the per capita land area in China.

The per capita arable land area in China is small, and the reserve resources of cultivated land are scarce. According to the statistics of the Ministry of Land and Natural Resources, China's total cultivated land decreased during 2001–2007. In comparison with 2001, the cultivated land area showed a net reduction of 0.88×10^8 mu (120 million ha) in 2008. However, new cultivated lands were added by land consolidation, reclamation, and new development in the same period, 4.0 %

more than the cultivated land used by construction. Compared with 2001, the occupation of cultivated land slowed. But cultivated lands still are disappearing.

According to the "Communique on Land and Resources of China 2008," published by the Ministry of Land and Natural Resources of China, China had 1.21×10^8 hm^2 of cultivated land (12.81 % of total area), 0.12×10^8 hm^2 of orchard land, 2.36×10^8 hm^2 of forest land, 2.62×10^8 hm^2 of pasture land, 0.25×10^8 hm^2 of land used for other agricultural uses, 0.27×10^8 hm^2 of land for residential and industrial/mining sites, 0.02×10^8 hm^2 of land for transport and communications, and 0.03×10^8 hm^2 of land for water conservation facilities. The rest was unused land.

There are significant regional differences of land productivity in China. China's water resources and arable land are not balanced between the northern and southern parts of the country, and this has created severe resource shortages in certain areas. There is more water but less arable land in the south and vice versa in the north. For example, arable land accounts for ~40 % in the North China Plain, a wheat- and cotton-producing area, while water resources in the region account for only ~6 % of the country's resources.

1.1.7 Oceans and Islands of China

China's mainland coastline is ~18,000 km long, extending from the Yalu River in the north to the Beilun Port in the south. The coastal areas are flat and contain many excellent harbors that are ice free all year. Eastern and southern China borders on the Bohai Sea, Yellow Sea, East China Sea, and South China Sea. China has a sea area of 4.73×10^6 km^2. The Bohai Sea is an inland sea, while Yellow Sea, East China Sea, and South China Sea are marginal seas to the Pacific Ocean.

The coastline of China is situated in several different uplift and subsidence zones of the Neocathaysian tectonic system. It also spans several different temperate, subtropical, and tropical climatic zones. Waves, tides, currents, and rivers have significant effects on the development of coasts, resulting in complex and diverse coastal types. In summary, the coastline of China can be broadly divided into three types: coastal plain, mountainous and hilly coast (bedrock coast), and biological coast. To the north of Hangzhou Bay, except for the mountainous and hilly coast in the Liaodong Peninsula and Shandong Peninsula, the vast majority of the coastline is coastal plain. To the south of Hangzhou Bay, except for some local harbors and coastal plains in the small- and medium-sized estuarine delta, the vast majority is mountainous and hilly coastline. Biological coast only exists in some parts along the coastal areas of the South China Sea and East China Sea.

China has more than 5400 islands. One of the biggest islands is Taiwan with an area of 36,000 km^2, followed by Hainan with an area of 34,000 km^2. The Diaoyu Islands and Akao, located northeast of Taiwan Island, are the easternmost islands of China. Islands and reefs scattered in the South China Sea are generally referred to as the South China Sea Islands, which is the southernmost island group of China. They are also known as the Dongsha Islands, Xisha Islands, Zhongsha Islands, and Nansha Islands.

1.1.8 Zonal Distribution of Land Characteristics and Geographical Regionalization in China

China covers a large area with vast latitude and longitude ranges and encompasses three types of zones: latitude zonality, longitude zonality, and elevation zonality.

Based on physical geography, precipitation and temperature, vegetation zones, and soil types, Zheng et al. (2008) partitioned China's eco-geographical regions including 11 temperature zones, 21 humidity regions, and 49 eco-geographical regions (Fig. 1.2; Table 1.2).

1.1.9 Natural Disasters in China

China is one of the countries with the most serious natural disasters in the world. With accelerating global climate change and China's rapid economic development and urbanization, there are increasing pressures on China's natural resources and environment, and subsequently more severe and complicated natural disasters. China has a very large territory and complex geographical and climatic conditions, with many kinds of frequently occurring natural disasters. Except for disasters

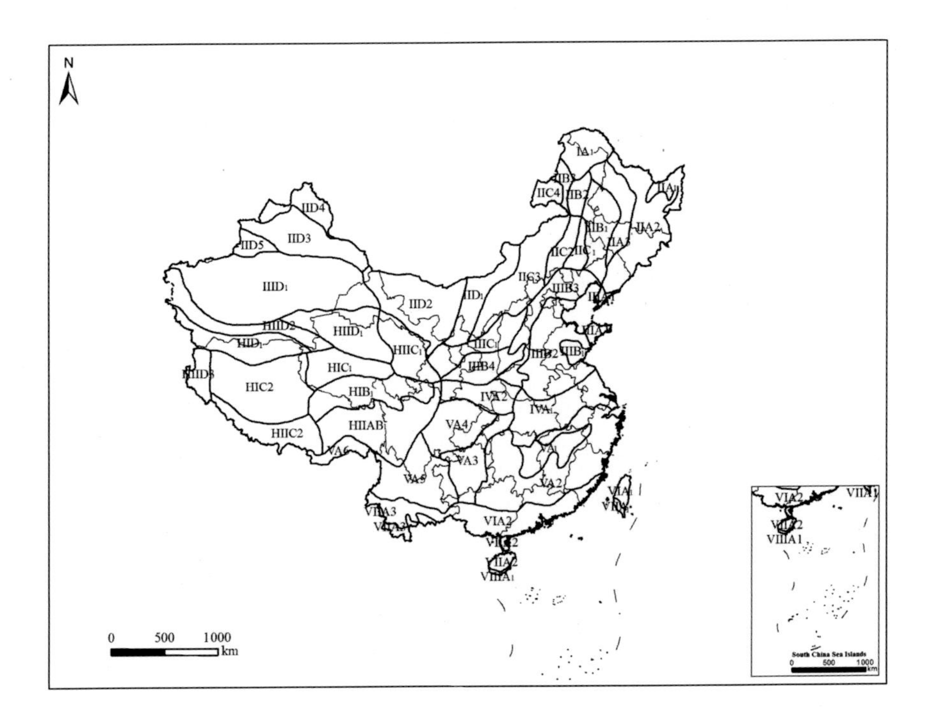

Fig. 1.2 Eco-geographical regions in China (see Table 1.2)

Table 1.2 China's eco-geographical region system

Zone	Humidity region	Eco-geographical region
I. Cold temperate zone	A. humid	I A_1. North Da Xing'an Ling Mountains deciduous coniferous forest region
II. Temperate zone	A. humid	II A_1. Sanjiang Plain wetland region II A_2. Xiao Xing'an Ling Mountains, Changbai Mountains, broad-leaved and coniferous forest region II A_3. Piedmont platform of eastern Song-Liao Plain broad-leaved and coniferous mixed forest region
	B. Subhumid	II B_1 Middle Song-Liao Plain forest steppe region II B_2 Middle Da Xing'an Ling Mountains steppe forest region II B_3 Hill land of north part of western Da Xing'an Ling Mountains piedmont forest steppe region
	C. Semiarid	II C_1 Western Liao River Plain steppe region II C_2 Southern Da Xing'an Ling Mountains steppe region II C_3 Eastern Inner Mongolia highland steppe region II C_4 Hulun Buir Plain steppe region
	D. Arid	II D_1 Ordos and Western Inner Mongolia Highland desert steppe region II D_2 Alax and Hexi Corridor desert region II D_3 Jungar Basin desert region II D_4 Altay Mountains steppe, coniferous forest region II D_5 Tianshan Mountains desert, steppe, and coniferous forest region
III. Warm temperate zone	A. Humid	III A_1 Eastern Liaoning and Jiaodong hill land deciduous broad-leaved forest and cultivated vegetation region
	B. Subhumid	III B_1 Middle Shandong hill land deciduous broad-leaved forest and cultivated vegetation region III B_2 North China Plain cultivated vegetation region III B_3 North China Mountains deciduous broad-leaved forest region III B_4 Fenhe and Weihe River basins deciduous broad-leaved forest and cultivated vegetation region
	C. Semiarid	III C_1 Northern and Middle Loess Plateau steppe region
	D. Arid	III D_1 Tarim basin desert region
IV. North subtropical zone	A. Humid	IV A_1 Plain of Changjiang River and Dabieshan Mountain evergreen and deciduous broad-leaved mixed forest and cultivated vegetation region IV A_2 Qinling and Bashan mountains evergreen and deciduous broad-leaved forest mixed region
V. Mid-subtropical zone	A. Humid	V A_1 Southern Changjiang River hill land evergreen forest and cultivated vegetation region V A_2 Fujian-Zhejiang and Nanling Mountains evergreen broad-leaved forest region V A_3 Hunan and Guizhou Mountains evergreen broad-leaved forest region

(continued)

Table 1.2 (continued)

Zone	Humidity region	Eco-geographical region
		V A_4 Sichuan Basin evergreen broad-leaved forest and cultivated vegetation region V A_5 Yunnan Plateau evergreen broad-leaved forest and pine forest region V A_6 Southern of East Himalayas Mountains seasonal rain forest and evergreen broad-leaved forest region
VI. South subtropical zone	A. Humid	VI A_1 Middle and Northern Taiwan Mountains and Plain evergreen broad-leaved forest and cultivated vegetation region VI A_2 Fujian, Guangdong, and Guangxi low Mountains and Plain evergreen broad-leaved forest and cultivated vegetation region VI A_3 Middle and Southern Yunnan Mountains and hill evergreen broad-leaved forest and pine forest region
VII. Marginal tropical zone	A. Humid	VII A_1 Southern Taiwan Mountains and Plain seasonal rain forest and rain forest region VII A_2 Hainan and Leizhou Mountains and hill land semi-evergreen seasonal rain forest region VII A_3 Xishuangbanna Mountains seasonal rain forest and rain forest region
VIII. Tropical zone	A. Humid	VIII A_1 South Hainan, Eastern and Middle Archipelagos seasonal rain forest and rain forest region
IX. Equatorial tropical zone	A. Humid	IX A_1 Nansha Archipelagos region
H I. Plateau subcold zone	B. Subhumid	H I B_1 Guoluo-Naqu Plateau Mountains alpine shrub meadow region
	C. Semiarid	H I C_1 Southern Qinghai Plateau and wide valley alpine meadow steppe region H I C_2 Qiangtang Plateau lake basin alpine steppe region
	D. Arid	H I D_1 Kunlun high mountain and plateau alpine desert region
H II. Plateau temperate zone	A/B. Humid/subhumid	H II A/B_1 Western Sichuan and eastern Tibet high mountain and deep valley coniferous forest region
	C. Semiarid	H II C_1 Qilian Mountains of eastern Qinghai high mountain and basin coniferous forest and steppe region H II C_2 Southern Tibet high mountain and valley shrub steppe region
	D. Arid	H II D_1 Qaidam Basin desert region H II D_2 North Kunlun Mountains desert region H II D_3 Ngali Mountains desert region

caused by modern volcanic activity, almost all types of natural disasters occur every year in China, such as floods, droughts, earthquakes, typhoons, hails, snowstorms, landslides, mudslides, pests and diseases, and forest fires.

According to the statistics of the Ministry of Civil Affairs and the National Disaster Reduction Committee in 2011, various types of natural disasters affected 430 million people (1126 people were killed including the 112 missing and 9.394 million people were evacuated), damaged about 32.471 million ha of crops, and destroyed about 935,000 houses, resulting in direct economic losses of 309.64 billion CNY.

In normal years, disasters affect ~200 million people, including deaths, relocation, destroyed crops, and collapsed houses. With the recent sustained and rapid development of the national economy, expanded production, and an accumulation of social wealth, there has been an increasing trend in losses. Natural disasters have become one of the main factors constraining the sustainability and stable development of China's economy. 74 % of provincial capital cities and more than 62 % of prefecture-level cities are located in dangerous areas with seismic intensities above VII. More than 70 % of large cities, with more than half of the population, and more than 75 % of the agricultural and industrial production of the country are in areas with serious meteorological, oceanographic, hydrographic, earthquake, and other hazards.

1.2 China's Climate

1.2.1 Basic Climate Characteristics

China's territory spans almost 50° in latitude, from 3°52′N to 53°31′N. China's topography varies from the Qinghai-Tibet Plateau at more than 4000 m above the sea level [peaked by Mt. Qomolangma (Everest) at 8848 m] to its east coastline on the Pacific Ocean. China's climate is governed by its geography and topography, with most of the territory falling in the monsoon zone.

China's climate shows global and regional characteristics. The most direct component influencing China's climate is the East Asian atmospheric circulation (monsoon). The basic airflows (monsoonal circulations) that govern China's climate differ significantly from winter to summer. In winter, China is mainly subject to the westerly winds over the Northern Hemisphere. In summer, China is dominated not only by the Indian low and the subtropical high (with southerly winds), but also by the combined impacts of the westerly circulation and regional airflow from the tropics.

At the surface, most parts of China are affected by both the southern branch of the cold Mongolian high and the Aleutian low in winter, with northwesterly, northerly, or northeasterly winds prevailing in the mainland.

In the middle troposphere (especially at the 500 hPa level), most parts of China is controlled by northwesterly or westerly airflows in winter. In contrast in summer, the areas near 40°N, and extending further north, are dominated by the westerly circulation, while the areas south of 40°N are governed by southeasterly or southerly or southwesterly airflows. The Qinghai-Tibet Plateau strengthens the monsoon circulation over eastern China that results from the land-sea difference, which makes winter and summer monsoons more significant. The Plateau in summer blocks the airflow directly, and prevents the summer monsoon from arriving at areas to its north, including Xinjiang, Gansu, and Inner Mongolia, where the climate is hot and dry. The Plateau in winter is a barrier that prevents the northern cold air from invading the south, and the middle- and low-level westerly flow from moving east. As a result, the invading cold air is forced to shift eastward, which leads to a warmer winter in areas to the southeast flank of the Plateau, including Yunnan, Guizhou, and Sichuan, and a colder winter in most areas of China than other places at the same latitude.

China spans multiple temperature zones: cold temperate, temperate, moderate temperate, warm temperate, northern subtropical, central subtropical, southern subtropical, marginal tropical, central tropical, and equatorial tropical belts, from north to south. It also covers some typical plateau climate zones (i.e., plateau frigid, temperate, and subtropical belts) and transcends various climate types (i.e., wet, humid, subhumid, semiarid, arid, and extremely arid, from southwest to northwest). Besides, monsoonal climate of China is characterized with the coincidence of dry and cold periods as well as rainy and warm weather in the same season, large difference between the south and the north in winter while small difference in summer, and distinctive advances or retreats of rain belts.

1.2.2 Temperature and Precipitation in China

China is generally cold in the north and warm in the south as well as cold on the plateaus and warm on the plains (Fig. 1.3). Based on in situ measurements from Chinese meteorological stations, the annual temperature averaged in mainland China is 8.8 °C, with temperatures decreasing from south to north. The difference in annual temperature between the south and the north is more than 30 °C, with more than 25 ° C registered in the Hainan Islands and −5 °C in the northernmost Mohe County. Similarly, the annual mean temperature gradually declines from the plains to the Qinghai-Tibet Plateau. Due to its topography, western China, including the plateau, registers an annual mean temperature below 0 °C. The annual mean temperature is 8 °C in the Tarim Basin as compared with 20 °C in eastern coastal regions of China.

January is the coldest month in mainland China, with temperatures dropping as low as −30 °C in the northern part of northeast China and as low as −30.9 °C in the northernmost Mohe region. The western alpine regions experience temperatures below −20 °C, such as in the Altai Mountains, central Tianshan Mountains, and northern Tibetan Plateau. Apart from northern and western China, eastern China

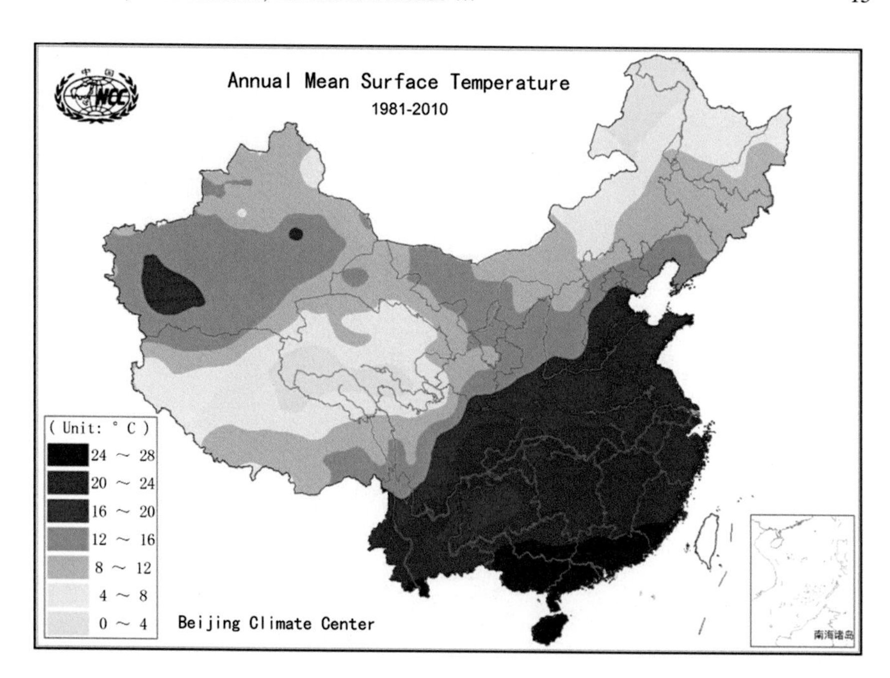

Fig. 1.3 Spatial distribution of China's annual mean surface temperature

has the coldest winters among locations at the same latitude in the world. This is because China is located in the low and middle latitudes of the Northern Hemisphere and is frequently affected by the Arctic cold air (the winter monsoon). The mean zero isotherm in January in China is close to 33°N, along the Qinling Mountains and the Huaihe River, which is the southernmost zero isotherm in the world. July is the warmest month in China, with the exception of the coastal areas and islands. The mean temperature is 18.4 °C at Mohe and nearly 30 °C in the south of the Huaihe River. Zhejiang and Shanghai frequently break records of continuous high-temperature days in consecutive years, and even Beijing (in the north) can experience high temperatures.

Precipitation distribution in China is characterized by more precipitation in the southeast and less precipitation in the northwest, with the isohyet declining from southeast to northwest (Fig. 1.4). According to the observational data from weather stations in China, the annual average precipitation in China is 612.9 mm, with the highest amount of 1500–2000 mm observed along the southeastern coast and southern Yunnan Province. Precipitation to the north of the Qinling Mountains, and the Huaihe River is generally 500–800 mm. The 400-mm isohyet tracks along the Da Xing'an Ling Mountains, Hohhot, and Lanzhou, up to the Yarlungzangbo River valley, and the isohyets below 100 mm are found in the areas to the west of the Helan Mountains. The lowest precipitation being only 5.9 mm per year is recorded at Turpan Tuokexun station. Under the monsoon climate, both hot and rainy days can occur in the same season.

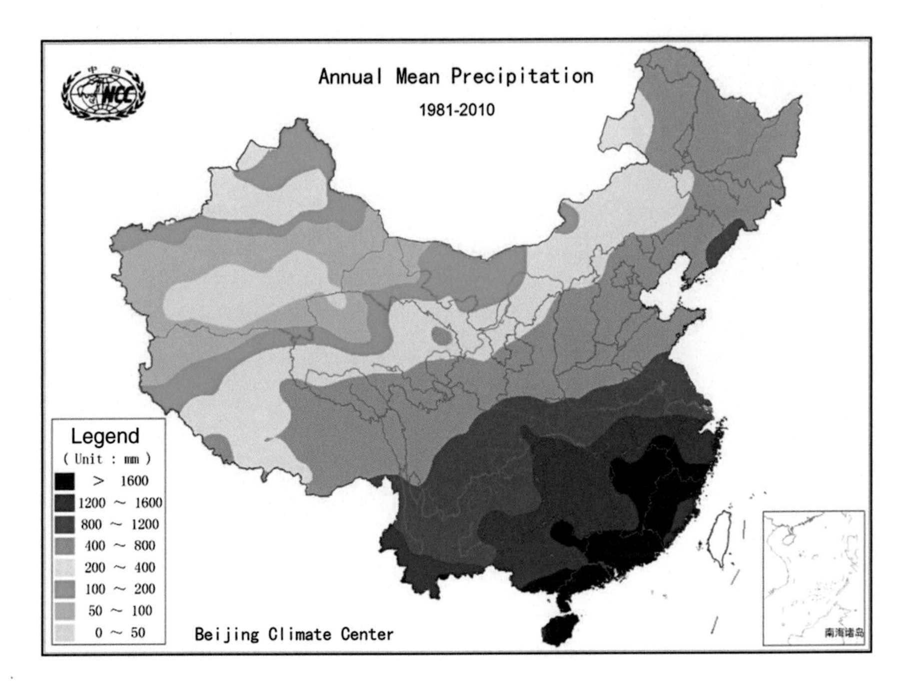

Fig. 1.4 Spatial distribution of China's annual mean precipitation

1.2.3 China's Climate Resources

With the exception of the northern temperate zone, the Qinghai-Tibet Plateau, and the alpine mountain regions, the rest of China has heat resources usable for agricultural production. Farming begins when the daily mean temperature is ≥0 °C, and the soil is thawing. The farming period in the northern part of northeast and northwest China is less than 200 days, compared with up to 300 days in the south of the Qinling Mountains and Huaihe River. The areas south of the Nanling Mountains experience a year-round farming period. The growing season length is defined as the period when the daily mean temperature is ≥5 °C for consecutive days. The growing season length is less than 150 days in the northern regions of northeast and northwest China, while 250 days in areas south of the Qinling Mountains and the Huaihe River. The growing season length for the areas south of 25°N reaches one whole year.

1.2.3.1 Solar Energy Resources

China is endowed with rich solar energy (Fig. 1.5). The annual average daily solar radiance in the vast majority of the country is above 4 kWh/m^2, with the highest radiance of 7 kWh/m^2 in the Tibet. The potential solar energy in China is totally 140 GWh/a, which is equivalent to 1.7 trillion tons of standard coal/a or 4038 times of China's total power generated in 2008.

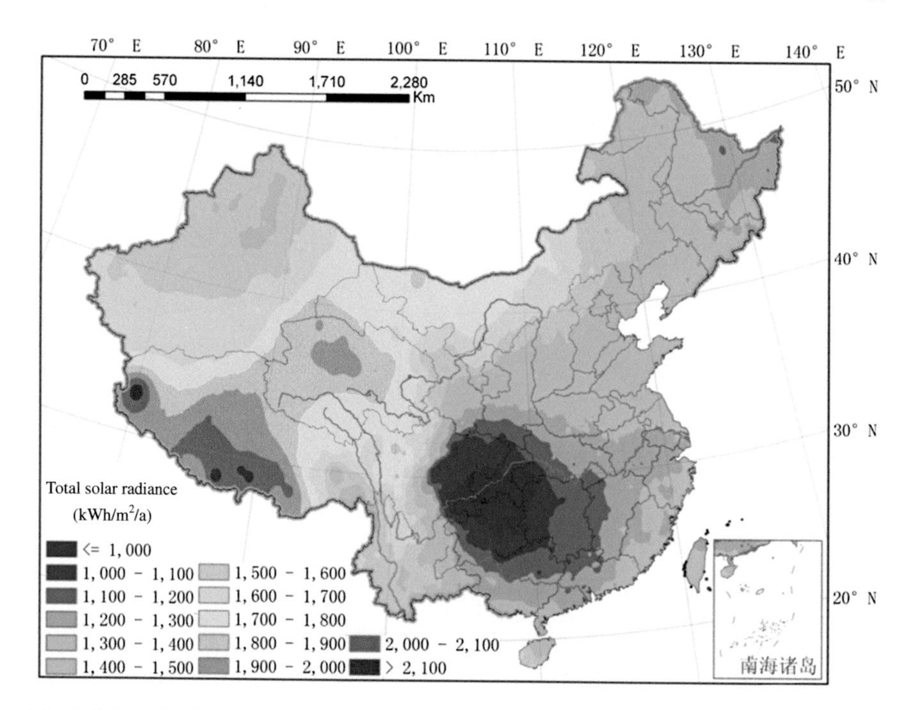

Fig. 1.5 Distribution of annual mean solar energy in China

1.2.3.2 Wind Energy Resources

Most of China is under a monsoon climate, while its northwestern regions experience a dry continental climate. The interactions between the cold air from westerly winds and the airflow from the Pacific Ocean appear frequently, offering China abundant wind energy resources in spite that they are unevenly distributed. The areas with abundant wind energy resources are mainly located in northwest China, north China, northeast China, and the eastern and southern coastal zones. Landforms like mountains and lakeshores in other regions also have abundant wind resources. Offshore wind energy resources are also rich in China, especially across the Taiwan Strait (the "canyon effect"), and in eastern part of Guangdong, the offshore region of Zhejiang, and the middle and north parts of the Bohai Bay. However, developing wind resources in the Taiwan Strait is difficult, because the area with water depths in the 5–50 m range is small. Although water depths are more favorable off the coast of Jiangsu, the wind is not strong in this area. The water depth in the entire Bohai Bay is less than 50 m (more than half is 5–25 m) and has huge wind energy development potential. However, typhoons and seasonal sea ice can adversely affect offshore wind energy development.

In 2011, the China Meteorological Administration (CMA) accomplished a detailed survey and evaluation on the national wind energy resources (Fig. 1.6). The results show that the technically exploitable amount of wind energy resources at 50,

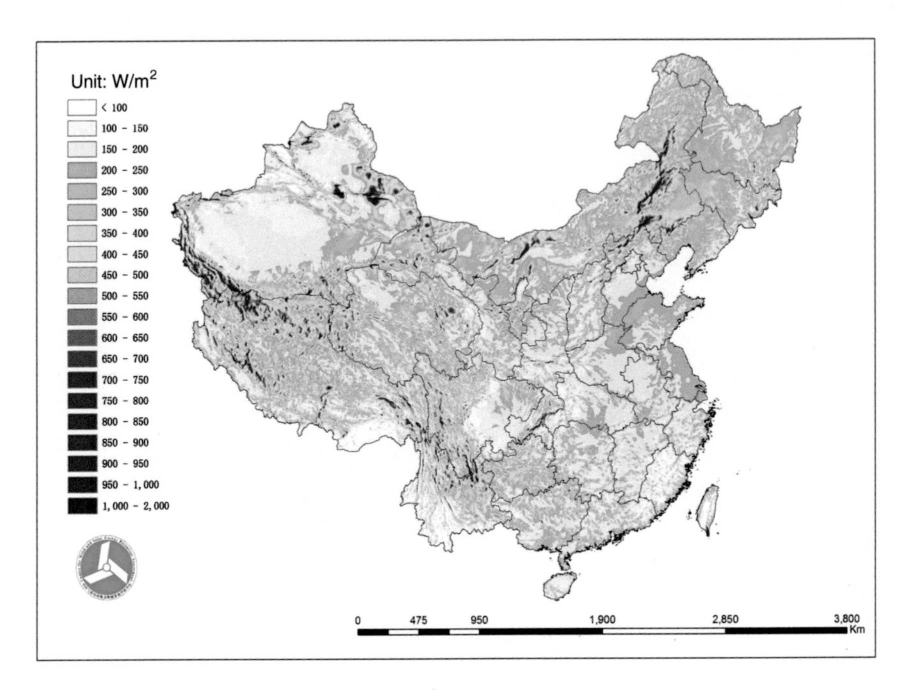

Fig. 1.6 Distribution of annual mean wind power density with 1 km horizontal resolution at 70 m height over mainland China

70, and 100 m above the mainland China (excluding the Qinghai-Tibet Plateau areas above 3500 m) were 2050 GW, 2570 GW, and 3370 GW, respectively (usable wind resources with installed capacity beyond 1500 kW/m^2, taking into account the major geographical and social factors that may restrict wind power development). 20 % of the offshore areas within 50 km of China's coasts can be used for wind energy deployment. It is estimated that the exploitable wind energy resources are ~510 GW at the height of 100 m above the water, if wind turbines are built in 5–50 m water depths.

1.2.3.3 Precipitable Water Resources

The total amount of annual mean precipitation over mainland China is 6.1889 trillion m^3, which is the source of surface runoff and river resources. 67.7 % of the precipitation is distributed in the south where rivers are abundant. However, precipitation across the country has monthly, seasonal, and interannual variations. For the whole country, the annual precipitation variability is between 10 and 50 %. The variability is low in the areas with more precipitation and high in the areas with less precipitation. For example, the precipitation variability is up to 30–50 % in the arid zones of northwest China. From the seasonal perspective, precipitation variability is low in the rainy season and high in the dry season. For example, the precipitation variability may reach 100 % in January in northwestern China. The water resources in China are also closely related to dryness (the ratio of maximum possible evaporation to precipitation in vegetated areas) and relative humidity.

1.2.4 Climate Zoning in China

China has a vast territory with unique geographical location, complex landforms, and diverse climate. Therefore, climate zoning provides a service for targeted industrial and agricultural activities, and socioeconomic development, and also a scientific basis for climate change adaptation.

In 1929, Professor Zhu Kezhen, a Chinese meteorologist, created the first climate zonation in China, with eight climate zones: south China, central China, north China, northeast China, Yunnan-Guizhou Plateau, Grassland, Tibet, and Mongolia-Xinjiang. In 1966, based on Zhu's and others' work, the Central Meteorological Service (now as China Meteorological Administration (CMA)) mapped China's climate zones using data from more than 600 stations across China in 1951–1960. This zoning was updated based on the latest climate data in 1978 and in 1994.

In 2010, based on daily weather observations from 609 stations nationwide, Zheng et al. (2013) rezoned China's climate into 12 temperature zones (i.e., northern, central, and warm temperate zones; northern, central, and southern subtropical zones; marginal, central, and equator tropical zones; plateau subcold, plateau temperate, and plateau subtropical zones), 24 arid/humid regions, and 56 climate zones. Compared with their predecessors' methods, zonation of Zheng et al. shows that since the 1970s, there have been no significant changes in the overall pattern of climate zones and regions in China, but some important climate boundaries shifted to some extent. The northern boundaries of the subtropical and warm temperate zones moved north, and the semihumid and semiarid boundaries in northern China shifted east and south to some extent, respectively. Changes also occurred with climatic subregions in the moderate, warm temperate, and north subtropical zones. These are probably due to warming in most of China and drought in some regions in northern China since the 1980s. In 2013, this zoning was updated based on the latest climate data during 1981–2010 (Fig. 1.7).

1.2.5 Weather and Climatic Disasters in China

Weather refers to a short timescale, and climate refers to a long timescale. For example, rainy season is a climate concept, and a heavy rainfall is a weather event. Usually, a rainy season lasts for one month or even longer and consists of multiple precipitation events (including heavy rainfall). Due to an excessively long (flood) or short (drought) rainy periods, or unusually little or abnormally much total rainfall,

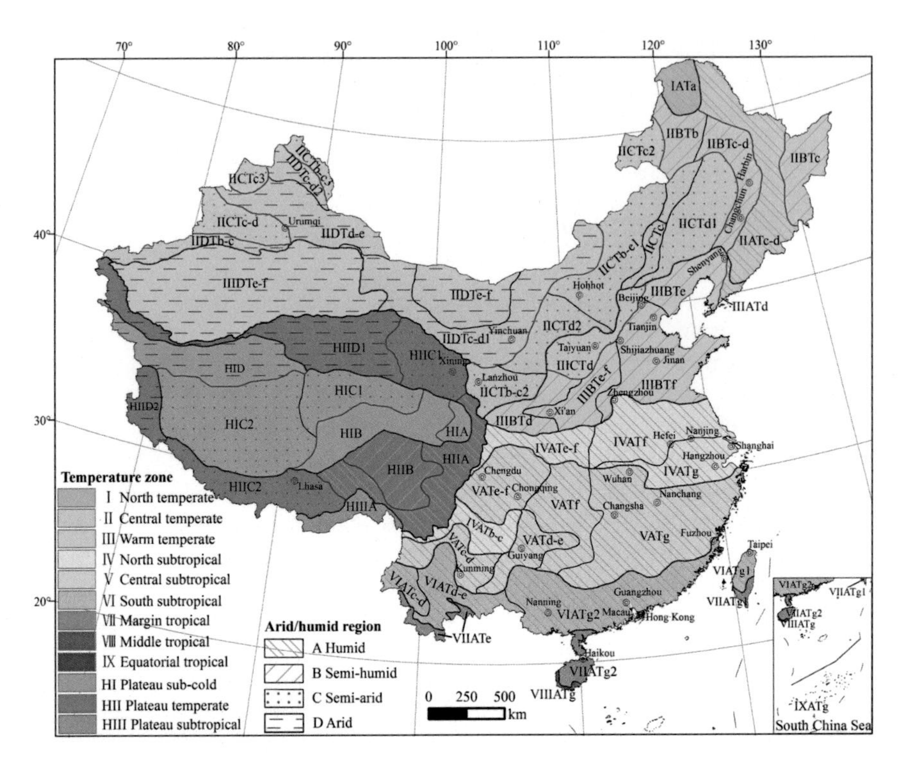

Fig. 1.7 Map of China's climate zones

drought or flood disaster may occur. In other words, climatic disasters refer to that caused by extensive, long, and persistent climate anomalies. Other examples include frequent cold waves in severe winters or few and weak cold waves in warm winters.

Because eastern China is located in the East Asian monsoon zone and western China are inland, China is susceptible to weather and climatic disasters. These disasters include typhoons, seawater inundation, thunderstorms, drought, heat waves, sand/dust storms, cold waves, gales, frost, snowstorms, hail, fog, haze, and acid rain. Climate-induced disasters in China show regional characteristics. In the eastern monsoon zone, winter can bring cold waves, gales, low temperatures, frosts and droughts, and summer can bring waterlogging, heat waves, droughts, and typhoons. Western China tends to be dry in one year-round, with strong winds and sand/dust storms in spring. This region, together with pasture lands on the Qinghai-Tibet Plateau, is subject to snowstorms. On the plateau, severe convection weather events such as gales, thunderstorms, and hail are common. Southwestern China, with complex landforms, is mainly exposed to droughts and rainstorms.

Since the 1990s, weather and climatic hazards have affected more than 48 million ha of cropland, on average, in China. Drought-affected areas are the largest, accounting for 49.8 % of the total disaster-affected area, followed by rainfall-induced floods over 26.1 % of the area, gales and hailstones in 10.4 %, freezing temperatures and heavy snow in 7.5 %, and typhoons in 6.2 %. Every year, meteorological disasters affect 380 million people and cause 4427 deaths, on average (in 1990, the most deadly year, the death toll was 7206). Annual average direct economic loss has reached 181 billion CNY, accounting for about 2.7 % of the nation's gross domestic product (GDP). In the most damaging year, 2010, the loss was recorded high of 509.75 billion CNY, accounting for 1.3 % of GDP (compared with 298.9 billion CNY in 1998). This was related to the exceptionally severe landslides and mudflow disasters in Zhouqu, Gansu Province.

Zhouqu is located in the southeastern part of Gannan Tibetan Autonomous State, covering an area of 3010 km^2, with a population of 134,700. In this region, the annual mean temperature is 12.7 °C, the annual average frost-free period is 223 days, and the annual precipitation is 400–800 mm. Being neither frigid in winter nor hot in summer, it is known as the "Orchard on Ridges." The county's topography is high in the northwest and low in the southeast; its landforms are complex, with many ravines and gullies extending in all directions. So it is highly vulnerable to disasters such as landslides and mudflows. Geological formation in the county provides a large quantity of debris when mud- and rock flows occur. In the evening of August 7, 2010, a sudden excessively heavy rain fell in the mountainous area to the northeast of the township, which lasted for 40 min with rainfall exceeding 90 mm and triggered mud- and rock flows.

1.3 China's Socioeconomic Characteristics

1.3.1 Population and Ethnic Groups

China has the largest population in the world, with 1.34 billion people in mainland China in 2010. Between 1949 and the 1980s, the population increased, except for the period from 1959 to 1961. The growth rate has slowed since the introduction of the family planning policy in the 1980s. At present, the key issue regarding China's population is the contradiction between the large population and low productive forces, causing a population imbalance.

This imbalance of population structure has a great impact on China's sustainable development. In the twentieth century, China's population began aging, and it now has the largest aging population in the world. It is predicted that the proportion of aging population may reach 30 % of the total population by the middle of the twenty-first century. China enters into this aging society with undeveloped and underdeveloped social and economic systems, which may bring about many challenges, such as labor shortages and other social burdens (aging population's pension, health care, and so on).

There are spatial differences of population in China. Population density is 450 people per square kilometer in east China, 250 per square kilometer in central China, and only 50 per square kilometer in west China. In the 1930s, Huanyong Hu proposed an Aihui-Tengchong line, known as the Huhuanyong population distribution line. This dividing line places southeast China with more than 94 % of the national population and 36 % of the total national territorial area and leaves northwest China with less than 6 % of the population and 64 % of the total national territorial area. The population density decreases from the southeast to the northwest and from coastal to inland areas. There is a significant increase of population in the urban zones along the coast and along rivers, and a significant reduction of population in regions such as Sichuan, Chongqing, Hubei, north Jiangsu, and central Henan. With 56 ethnic groups, China is a multiethnic nation. In 2010, 8.49 % of the national population was ethnic minority, or 113.79 million people. There is a wide ethnic minority population distribution, mainly in the western and border areas of China, especially in Guangxi, Yunnan, Guizhou, and Xinjiang, which have more than half of the ethnic minority population. Due to historical and geographical factors, there is a wide gap of population density between ethnic minority areas and ethnic Han areas.

At present, the population growth rate is much larger in rural and poverty-stricken area, and in ecologically vulnerable area, than in urban and relatively developed regions. There has been negative population growth in Beijing, Shanghai, Tianjin, and other cities, while the population in northwest China continues to grow. The population growth rate is much higher than the national average level in the Taihang Mountains area, Qinba Mountains area, Dabie Mountains area, Dingxi arid mountain area, and Xihaigu area. In addition, these regions experience many problems such as poverty and low quality of the living and interlaced with complex ecological conflicts.

With impact of future climate change, stable food yields in China will become difficult, and economic activities such as high-energy consumption and environmental pollution will have to be curtailed. The country will also have to deal with the imbalance of economic growth and income distribution, and with urban–rural coordination. With industrialization and urbanization, there are many significant and deep-rooted population issues, such as landless peasants, unemployment, and poverty. These will have a significant impact on future sustainable growth and addressing climate change in China.

1.3.2 Economic Development

As one of the world's emerging economies, China has experienced rapid and continuous economic growth since the introduction of reform and opening-up policy. In 2010, the national GDP reached over 40.12 trillion CNY, ranking second in the world. Primary industry output was 4.05 trillion CNY (10.1 %), secondary industry output was more than 18.76 trillion CNY (46.8 %), and tertiary industry output was 17.31 trillion CNY (43.1 %) (Fig. 1.8). National financial revenue and investment in fixed assets for the whole society have both increased, and reached over 8.31 trillion CNY and 27.814 trillion CNY in 2010, respectively.

With continuous reforms and policies favoring open trade, there has been a rapid development of foreign trade and economic cooperation in China. The total value of imports and exports, and foreign investment reached over 2.97 trillion US dollars and 108.82 billion US dollars in 2010, respectively. The total direct foreign investment was more than 1.25 trillion US dollars from 1979 to 2010 (Fig. 1.9). The foreign exchange reserves of China grew steadily and reached over 2.85 trillion US dollars in 2010. However, the rapid economic growth of China depends mainly on export and investment. China has an economic structure of processing export orientation, and an industrial structure with a high proportion of the basic raw material

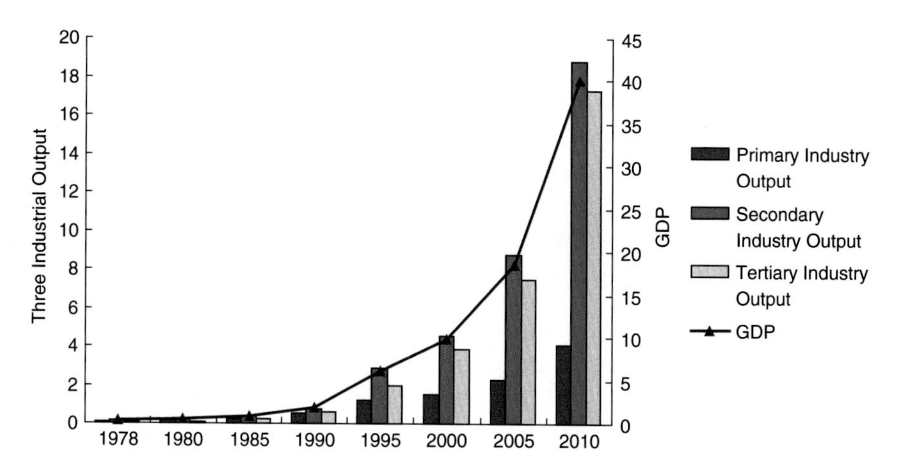

Fig. 1.8 Economic growth in China since 1978 (trillion CNY). *Note* Data are at current prices, sourced from China's Statistical Yearbooks, various years

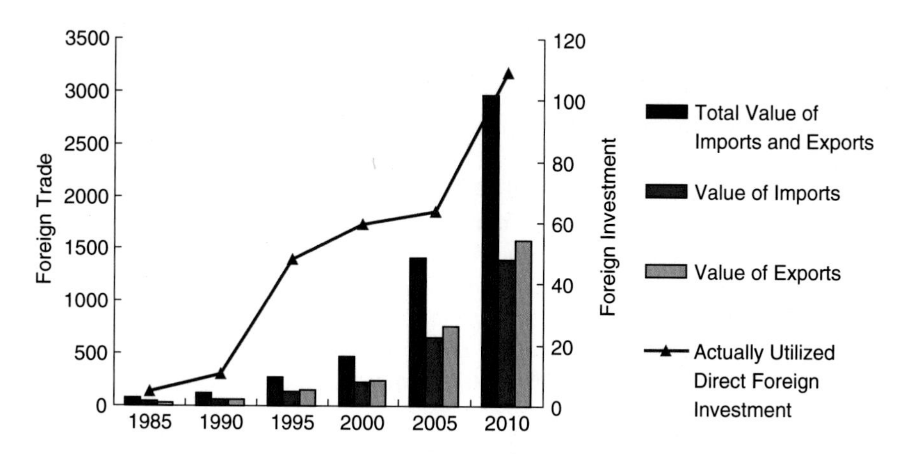

Fig. 1.9 Foreign trade and economic cooperation development in China (billion US dollars). *Note* Data are at current prices, sourced from China's Statistical Yearbooks, various years

industry and high-energy-consumption industry, which is the main determinant for the rapid increase in China's carbon dioxide emissions. Carbon dioxide emissions in exported goods represented 30–35 % of China's annual total emissions (Zhang 2009).

The rapid economic growth in China is based on an extensive growth mode and has a large dependence on fossil energy consumption. The energy consumption of China was less than half of that of the USA in 1999, but was equal to that of the USA in 2009. The consumption of primary energy has been increasing continuously since 2008 and reached more than 3 billion tons of standard coal equivalents (SCEs) in 2010, with an annual growth rate of 8 %. China has been the largest consumer of energy in the world, while the energy consumption of the USA and European Union has been steadily decreasing. The energy consumption of India is also increasing. Since the beginning of the twenty-first century, the gap has been expanding between energy production and consumption in China, and China's energy demand dependency will continue to increase on the world's energy supply. China's dependency on foreign oil was more than 55 % in 2011.

With rapid economic growth, carbon dioxide emissions have been increasing rapidly in China, accounting for 5.7 % of global emissions 1971, 10.7 % in 1990, and 23 % in 2005 (the highest emissions in the world; US emissions were only 16 % at that time). In the rest of the world's developed countries, carbon dioxide emissions are being reduced through new energy sources and the promotion of energy efficiency. However, China's carbon dioxide emissions continue to increase and could peak in 2030. With a 1 % increase in real GDP per capita, carbon dioxide emissions will increase by about 0.41–0.43 % (Li et al. 2011). China is confronted with two great challenges: addressing climate change in the international arena and protecting its environment during a domestic economic transition.

Regional development disparity is an increasingly serious problem in China, which puts a lot of constraints on mitigating and adapting to climate change. In 2010, the GDP and total export–import value in east China accounted for 53.1 and 87.6 % of the total national levels, respectively, while the region makes up 38 % of the national population and 9.5 % of the total land area. However, GDP and total export–import value in west China accounted for only 18.6 and 4.3 % of the total, respectively, and that region has 27 % of the national population and 71.5 % of the total land area.

China has recently experienced rapid and stable economic development, industrialization, urbanization, and further internationalization. The national comprehensive strength has been significantly enhanced, and people have enjoyed a relatively high living standard, which has given a boost to China's international status and influence. Tremendous achievements have been made in the construction of a socialist economy, socialist culture, and ecological civilization. China's carbon dioxide emissions per unit of GDP in 2020 could decrease to 40–45 % of those in 2005 based on the implementation of current projects. However, China is in the primary stage of socialism that will remain so for a long time. Many imbalanced and non-sustainable elements could turn into a tightened constraint on economic growth, including natural resources and environment conditions, unreasonable industrial structure, gap between regional and rural–urban development, and some institutional obstacles.

1.3.3 Urbanization

The urbanization rate of China was only 19.39 % in 1980, but this accelerated after the 1980s, and the urban population reached over 669 million people in 2010, 49.9 % of the total population. Urbanization in China is undergoing rapid development, in spite of meeting many challenges such as a limited supply of urban infrastructure and energy, environmental protection, and carbon dioxide emission reduction. Based on prediction, China's urbanization rate could reach 58–63 % in 2020 and 80 % by 2050.

Urbanization is unbalanced between different regions in China. Cities in east China accounted for 43.5 % of the total number, with 40 % of the urban; however, west China has only 19.1 % of all cities, with 71.5 % of the total national land area. The urbanization rate of west China is far less than the national average level. In addition, the spatial differentiation in city size in China is obvious. East China has more than 56 % of the total number of megacities, over 57 % of big cities, and about 50 % of medium-sized cities, and west China has a large proportion of small cities.

At present, GDP, energy consumption, carbon dioxide emissions, and sulfur dioxide emissions in China's urban region account, respectively, for 85, 85, 90, and 98 % of national total amount. Carbon dioxide emissions of more than 287 cities in China account for 72 % of total emissions. The contradiction between China's urban development and low-carbon transformation has become increasingly prominent, and there are many challenges in energy conservation and environmental improvement with the ongoing urbanization process.

The urban agglomerations (megacities) play an important role in China's socioeconomic development. At present, 10 urban megacities have been shaped in China, among which are the three national first-level megacities in the Yangtze River Delta, in the Pearl River Delta, and in the Beijing-Tianjin-Hebei area in the coastal region, and seven sublevel megacities in the Shandong peninsula, central-south Liaoning, central Henan Plain, Wuhan, Chang-Zhu-Tan, Chengdu-chongqing, and central Shaanxi Plain. With China's huge population size and rapid population growth, there will be a more rapid urbanization process in the future, and more and more people and industries will aggregate into the cities. New megacites may shape in the Harbin-Daqing-Qiqihar area, west bank of Taiwan Strait zone, surrounding zone of Poyang Lake, central Yunnan, Guangxi North Bay, Lanzhou-Xining-Germu, and the northern slope region of Tianshan Mountains. The National 12th Five-Year Plan proposed taking metropolis or large cities as bracing, and small cities and medium-sized cities as keynote, and shaping several significant urban agglomerations. However, climate changes and urbanization are two main factors making human beings more vulnerable to disasters, and the two factors superimpose with each other in urban agglomeration. Furthermore, population, industry, and infrastructure are highly centralized in urban agglomerations, which would become the high-risk areas being vulnerable to disasters and easily suffering great losses. The environmental and disaster problems presented by future urban agglomeration may become increasingly complicated and serious with climate change (Dong et al. 2012).

1.3.4 Social Development

China has recently experienced increasing social development, with an accompanying improvement in people's well-being. The annual per capita disposable income of urban households and annual per capita net income of rural households are growing steadily, reaching 19,109 CNY and 5919 CNY in 2010, respectively. The Engel's coefficients of urban and rural households were reduced to 35.7 and 41.1 %, respectively (Fig. 1.10).

The implementation of active employment efforts is being continuously strengthened in China, and the level of employment is rising. The social security system is also being bolstered and expanded. In 2010, 11.68 million newly found employments were created in urban areas, and the urban registered unemployment rate was only 4.1 %. At the end of 2010, the number of urban residents covered by basic old-age insurance, basic medical insurance, unemployment insurance, workers' compensation, and maternity insurance rose substantially to 256.73 million, 432.06 million, 133.76 million, 161.73 million, and 123.06 million, respectively.

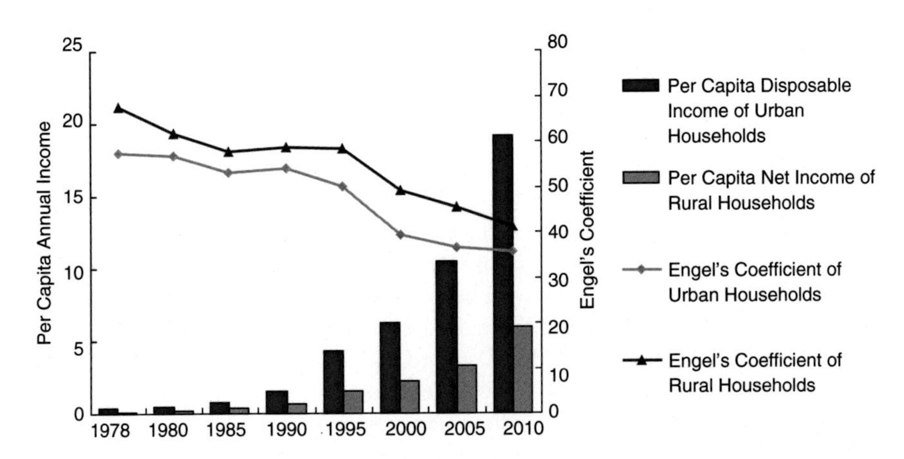

Fig. 1.10 Per capita annual income and Engel's coefficients of urban and rural households in China (thousand CNY, %). *Note* Data are at current prices, sourced from China's Statistical Yearbooks, various years

China has increased its development in the areas of science, education, culture, health, and sports. China's scientific and technological innovation capabilities continue to be enhanced. In 2010, there were 2358 regular institutions of university and college, with a postgraduate education enrollment of 0.54 million. China's research and development expenditures reached 706.26 billion CNY in 2010, accounting for 1.76 % of GDP. Acceptance of new patent application reached 1.22 million during that same time, and the transaction value in the technical market was 39.07 million CNY in 2010. Reforms and development of medical and health services are being actively carried out. At the end of 2010, there were 0.94 million medical and health institutions in China, with 5.88 million medical and health professional. The infrastructure of public cultural and sports has been greatly strengthened. By 2010, there were 0.31 million institutions in culture and cultural relics, 2.11 million employees, and 7159 sports institutions with employees of 0.16 million.

Overall, China has achieved remarkable progress in many social services and its people's living standards, and it has laid a sound foundation for mitigating and adapting to climate change. However, it still faces many challenges in social development. It has become not only more urgent but also more difficult to solve institutional and structural problems, and to alleviate the problem of imbalanced, uncoordinated, and unsustainable development, such as a relatively large income disparity, uncompetitive technological innovation capability, a coexistence of employment pressure and structural contradiction, and a significant increase in social conflicts.

References

Chen, B., & Li, J. (2008). Characteristics of spatial and temporal variation of seasonal and short-term frozen soil in China in recent 50 years. *Chinese Journal of Atmospheric Sciences, 32*(3), 432–433. (in Chinese).

Dong, S. C., Tao, S., Yang, W. Z., et al. (2012). Impact of climate change on urban agglomerations in coastal region of China. *Chinese Journal of Population, Resources and Environment, 10*(2), 78–83.

Li, F., Dong, S. C., Li, X., et al. (2011). Energy consumption-economic growth relationship and carbon dioxide emissions in China. *Energy Policy, 39*(2), 568–574.

Liu, C. H., Shi, Y. F., Wang, Z. T., et al. (2000). Glacier resources and their distributive characteristics in China—A review on Chinese glacier inventory. *Journal of Glaciology and Geocryology, 22*(2), 106–112. (in Chinese).

Zhang, X. P. (2009). Carbon dioxide emissions embodied in China's foreign trade. *Acta Geographica Sinica, 64*(2), 234–242. (in Chinese).

Zheng, D., Yang, Q. Y., Wu, S. H., et al. (2008) Research on eco-geographical region systems of China. Beijing: The Commercial Press. (in Chinese).

Zheng, J., Bian, J., Ge, Q., et al. (2013). The climate regionalization in China for 1981–2010. *Chinese Science Bulletin (Chinese Version), 58*(30), 3088–3099. (in Chinese).

Chapter 2
Climatic and Environmental Changes in China

Yong Luo, Dahe Qin, Renhe Zhang, Shaowu Wang and De'er Zhang

Abstract This chapter addresses climate change of instrumental era in China, including changes in the distribution and regional characteristics of the temperature, precipitation, Asian monsoon, general circulations, extreme weather and climate events, cryosphere (glaciers, frozen ground, and snow cover), sea-level rise, sea surface temperature (SST), and salinity. It also assesses climate variations on

---

The following researchers have also contributed to this chapter

Jiawen Ren

State Key Laboratory of Cryospheric Sciences, Cold and Arid Regions Environmental and Engineering Research Institute, Chinese Academy of Sciences, Lanzhou, China
e-mail: jwren@lzb.ac.cn

Tingjun Zhang

College of Earth and Environmental Sciences, Lanzhou University, Lanzhou, China
e-mail: tjzhang@lzu.edu.cn

Panmao Zhai

Chinese Academy of Meteorological Sciences, China Meteorological Administration, Beijing, China
e-mail: pmzhai@cma.gov.cn

Y. Luo (✉)
Center for Earth System Science, Tsinghua University, Beijing 100084, China
e-mail: yongluo@tsinghua.edu.cn

D. Qin
State Key Laboratory of Cryospheric Sciences, Cold and Arid Regions Environmental and Engineering Research Institute, Chinese Academy of Sciences, Lanzhou 730000, China
e-mail: qdh@cma.gov.cn

R. Zhang
Chinese Academy of Meteorological Sciences, Beijing 100081, China
e-mail: renhe@cams.cma.gov.cn

S. Wang
Department of Atmospheric and Oceanic Sciences, School of Physics, Peking University, Beijing 100871, China
e-mail: swwang@pku.edu.cn

D. Qin et al. (eds.), *Climate and Environmental Change in China: 1951–2012*,
Springer Environmental Science and Engineering,
DOI 10.1007/978-3-662-48482-1_2

different timescales (130, 20, 10, 2, and 0.5 ka) based on proxy archives such as sediments, ice cores, tree rings, and historical documents. Lastly, the advances in numerical paleoclimate simulation are summarized.

Keywords Climate change · Instrumental observation · Paleoclimate change · Numerical simulation

2.1 Climate Change in China in the Past 100 years

Measurements of climate variables in China began in the 1880s. Since then, these measurements have revealed significant climate warming, particularly since the 1980s.

2.1.1 Temperature Variations in China

It is difficult to establish a long time series of average temperatures in China because of a lack of systematic observations before the mid-twentieth century. Particularly, there have been few meteorological observations in inland of China. Currently, three surface air temperature series, extending more than 100 years, have been reconstructed for China. The temperature series W was calculated by integrating instrumental measurements with proxy data, such as ice cores, tree rings, and historical documents (updated from Wang et al. 1998). The instrumental data come from 50 meteorological stations, with 5 stations in each of 10 regions of China, including Xinjiang, Tibet, and Taiwan. The temperature series T was calculated similar to series W, but applying no proxy data (updated from Tang and Ren 2005). The temperature series C was created with Climatic Research Unit (CRU) datasets, in which data deficiencies over western China are filled in by interpolation from adjacent foreign observations (updated from Wen et al. 2006). High correlation coefficients, ranging from 0.78 to 0.93, among these three temperature series show good consistency. Series W and series T show a warming trend in China during 1906–2005 of 0.53 °C per 100 year (series W) and 0.86 °C per 100 year (series T). Overall, China has warmed by 0.5–0.8 °C during the past 100 years.

Instrumental meteorological observations have been conducted in China with a better coverage since A.D. 1951, at a maximum of 2200 stations. The temperature variations documented by these observations closely correlate with the series W and T data, showing that average surface air temperature has increased 1.38 °C between 1951 and 2009, with a trend of 0.23 °C per 10 year (Fig. 2.1).

D. Zhang
National Climate Center, China Meteorological Administration, Beijing 100081, China
e-mail: derzhang@cma.gov.cn

D. Qin
China Meteorological Administration, Beijing 100081, China

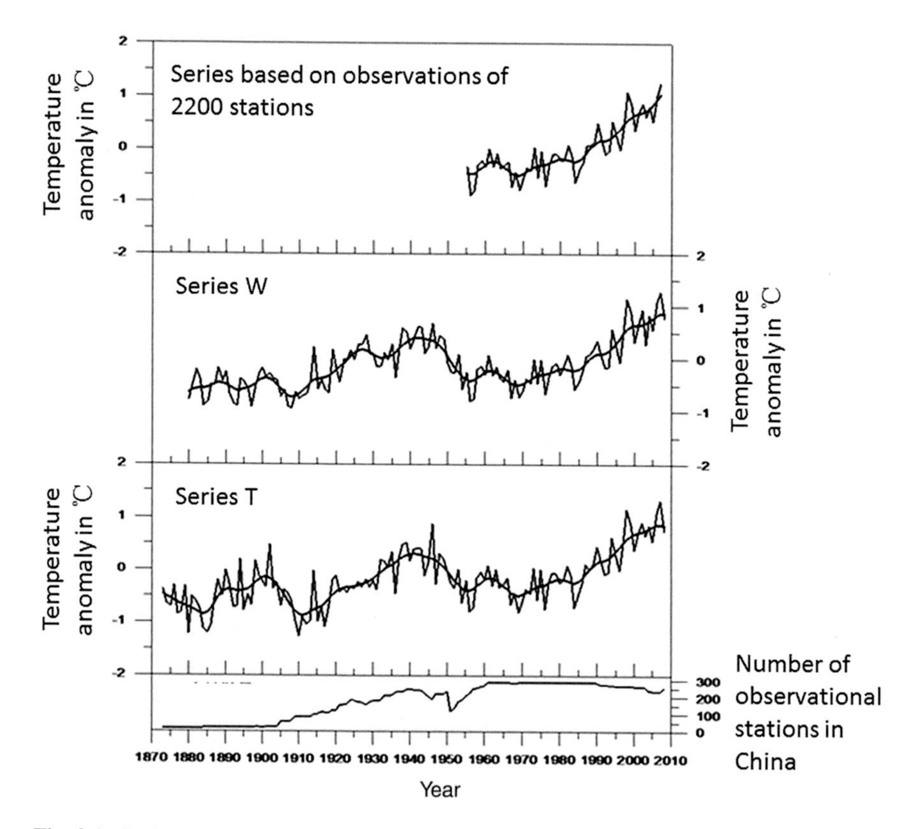

Fig. 2.1 Surface air temperature anomalies across China in 1873–2008 (relative to 1971–2000). The upper series is based on observations of 2200 stations; the middle two are series W and series T; the bottom panel shows the number of observational stations in China applied in series T

2.1.2 *Precipitation Variations in China*

Four precipitation datasets exist for China. The first dataset, starting from A.D. 1880, is based on seasonal total precipitation anomaly data from 71 stations in eastern China (updated from Wang et al. 2000). The second dataset, beginning from 1901, was constructed using monthly 1°×1° precipitation grid data from the CRU (updated from Wen et al. 2006). The third dataset uses data from 1951 onward and is based on the monthly dataset from 160 stations issued by National Climate Center, China Meteorological Administration. The fourth dataset set up by National Climate Center is similar to the third one except that it is based on observations from 2200 stations. Among these four precipitation datasets, the first has the longest time series, but it excludes western China, while the second has significant uncertainties in the early time series due to data deficiencies over western China. Both the third and fourth datasets cover all of China, but the observations start in 1951. Although these four datasets are calculated based on different source data with different methods, they show consistency in the common period, with correlation coefficients reaching 0.84–0.91. The data show that during the past 50–100 years, precipitation across China varied on a cycle of 20–30 years, with no significant trend compared with surface air temperature. Additionally, the precipitation variations have intensified temporal and spatial variability; for instance, there was excessive precipitation over the lower middle reaches of the Yangtze River but less precipitation over both north China and south China during the 1980s, as a reversal of the pattern in the 1970s.

2.1.3 Variations of Other Climate Parameters

Cloud cover plays an important role in Earth's radiation energy budget and hydrologic cycle. Cloud cover is currently measured with ground-based observations and with satellite observations. Satellite observations have a global coverage, but with shorter time span. In contrast, ground-based observations have been conducted for longer time periods, but they are less accurate than satellite data. Both data from ground and satellite observations show a decreasing cloud cover over China since 1961, especially in north China.

Solar radiation reaching the Earth surface has decreased by 2.5 % per decade during the past five decades, but there is a slight increase since the 1990s.

Upper air temperatures have also been changing over the past 50 years. Data show rising temperatures in the lower troposphere, but with a smaller warming amplitudes than that of surface air temperatures. In contrast, the temperature has a decreasing trend in the upper tropospheric temperatures, with a more pronounced decrease in stratospheric temperatures.

2.1.4 Variations in the East Asian Monsoon

China is located in the region of the East Asian monsoon (EAM), with most of China being influenced by the EAM, especially eastern China. The onset of summer EAM usually occurs in the South China Sea in mid-May. The raining season over the Yangtze valley (Meiyu) begins in mid-June and ends in mid-July, when summer EAM migrates to the north China. The winter EAM usually begins in October in China and slowly diminishes northward, ending in May of the next year. Both the summer and winter EAM have significant interannual and interdecadal variability. The winter EAM was strong from the 1920s to the early 1980s, but has weakened since the 1980s. The summer EAM was strong from the beginning of twentieth century to the end of the 1960s, with the strongest intensity in the 1960s. Since 1980, it has weakened steadily (Fig. 2.2).

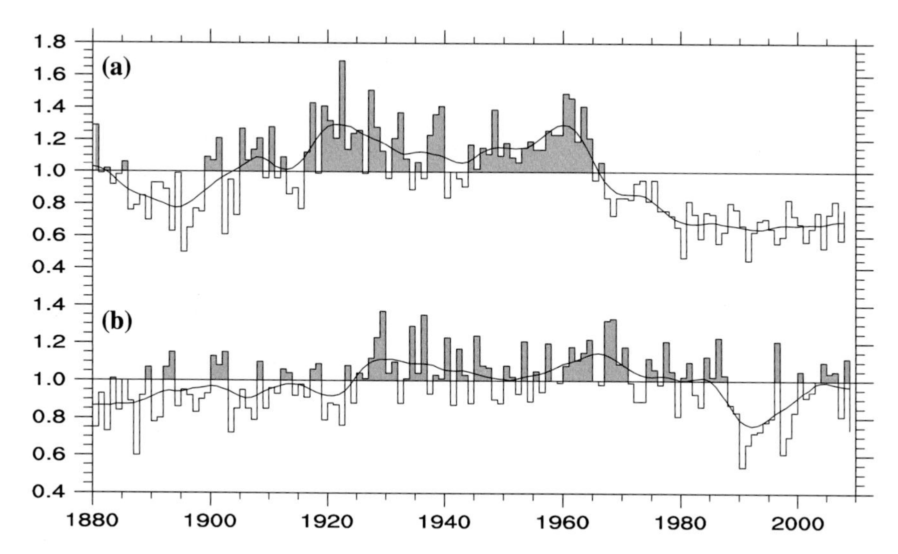

Fig. 2.2 Intensity of summer (**a**) and winter (**b**) EAM during 1880–2009. The East Asian monsoon intensity index is updated from Guo et al. (2003)

2.1.5 Atmospheric Circulation in East Asia

Atmospheric general circulations can affect the climate of China, among which are three prominent high-pressure systems: the Blocking High, the Subtropical High, and the South Asian High. The Blocking High is a high-pressure system at the geopotential height of 500 hPa (middle troposphere) in northeastern East Asia, i.e., the Okhotsk High. Sometimes, another Blocking High exists over the Ural Mountains. In particular, when low-pressure system occurs over Lake Baikal between the two high-pressure systems to its east and west sides in summer, this circulation pattern is typical in Meiyu season. The Subtropical High refers to the high-pressure system over the western Pacific Ocean at the geopotential height of 500 hPa. Its position and intensity play a decisive role in the position of the summertime rain belt in China. The South Asian High refers to the high-pressure system over the southern portion of the Eurasian continent at the geopotential height of 100 hPa (upper troposphere). It can also influence the position of the rain belt in China in summer, due to its close relationship with the Subtropical High over the western Pacific. The intensity of these three high-pressure systems has increased during the second half of twentieth century.

2.2 Variations in Extreme Weather and Climate Events

Climate warming may affect the intensity and frequency of some extreme weather and climate events.

2.2.1 Extreme Maximum and Minimum Temperatures

The annual averaged maximum surface air temperature in China has increased by 0.1 °C per decade. The number of days with a daily maximum temperature above 35 °C experiences significant interdecadal variations. There were more extremely hot days during the period from the 1960s to the early 1970s and in the past 10 years than that from mid-1970s to mid-1980s. This trend shows that the number of extremely hot days has risen gradually since the end of the 1980s (Fig. 2.3). The annual averaged minimum temperature in China is rising at a rate of 0.3 °C per decade (Fig. 2.4). Meanwhile, the averaged frequency of cold wave invasion in China has decreased by 0.3 per decade, and the number of days of frost injury has also decreased.

2.2.2 Extreme Precipitation

Since 1961, the frequency of regional severe precipitation events has slightly increased, with obvious interdecadal variations. The maximum number of severe precipitation events occurred in 1995 (14 events), and the minimum occurred in 1988 (2 events). The severe precipitation events were more frequent from the end of 1980s to 1990s.

The spatial distribution of extreme precipitation events (EPE) is complex. EPE over the Yangtze River and to the south of the Yangtze River tended to be stronger and more frequent, with an opposite trend over northern China. There was no obvious change in the averaged intensity of EPE over the west portion of northwest China, but the frequency of EPE tended to increase.

The number of continuous rain days in western China is increasing and decreasing in eastern China. The number of continuous rain days decreased significantly over northern, northeastern, and southeastern China and increased over the eastern Tibetan Plateau and in some areas along the southeast coast of the country.

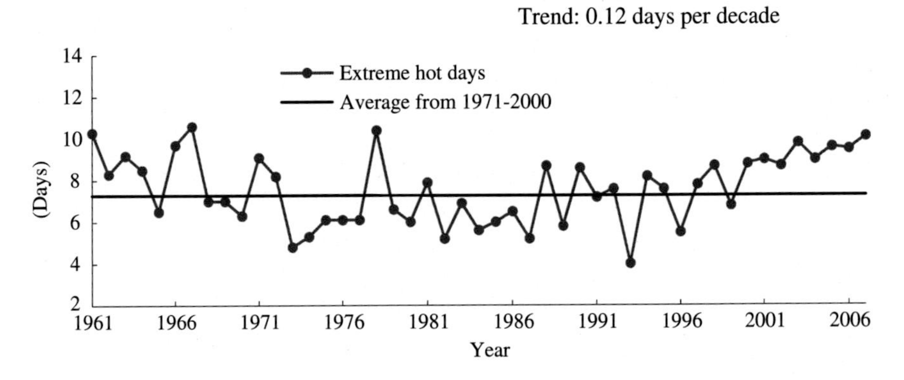

Fig. 2.3 Variation of the annual averaged extreme hot days during 1961–2007 in China, provided by National Climate Center

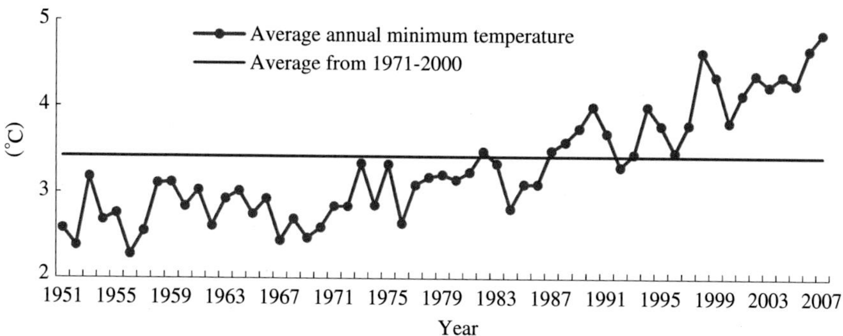

Fig. 2.4 Variation of the annual averaged minimum air temperature during 1961–2007 in China, provided by National Climate Center

2.2.3 *Drought*

Significant droughts have occurred in the eastern part of northwest China, in north China, and in northeast China over the past 50 years, while severe droughts have decreased in the region to the south of the Yangtze River and in the western part of northwest China (Fig. 2.5). Data show an obvious interdecadal variation in droughts in most parts of south China over the past 50 years. The number of continuous days without precipitation has been increasing in northern China for the past 50 summers. North China experienced an extensive drought for four consecutive years during 1999–2002 after 1997. In the summer of 2006, a severe drought occurred in Chongqing, a southwest city in China, with the return period of one hundred years. And the most severe drought during 50 years occurred in Sichuan Province in that year. Drought hit southwest five provinces, including Yunnan in 2010. Both the amount of precipitation and the number of rainy days decreased significantly in the regions. The surface air temperature increased significantly over northern China, and this increase in temperature enhanced the potential soil evaporation and thus intensified the droughts over the whole northern China.

2.2.4 *Tropical Cyclones*

From 1951 to 2011, the frequency of typhoons (tropical cyclones with winds ≥8 on the Beaufort scale) that were generated in the northwest Pacific Ocean and South China Sea has decreased with obvious interdecadal variations. And typhoon activities have been less than normal since 1995 (Fig. 2.6).

The frequency of typhoons making landfall in China did not change significantly over the same period, but there was great interannual variability. The maximum number of typhoons striking China was 12 (in 1971), and the minimum was 3 (in

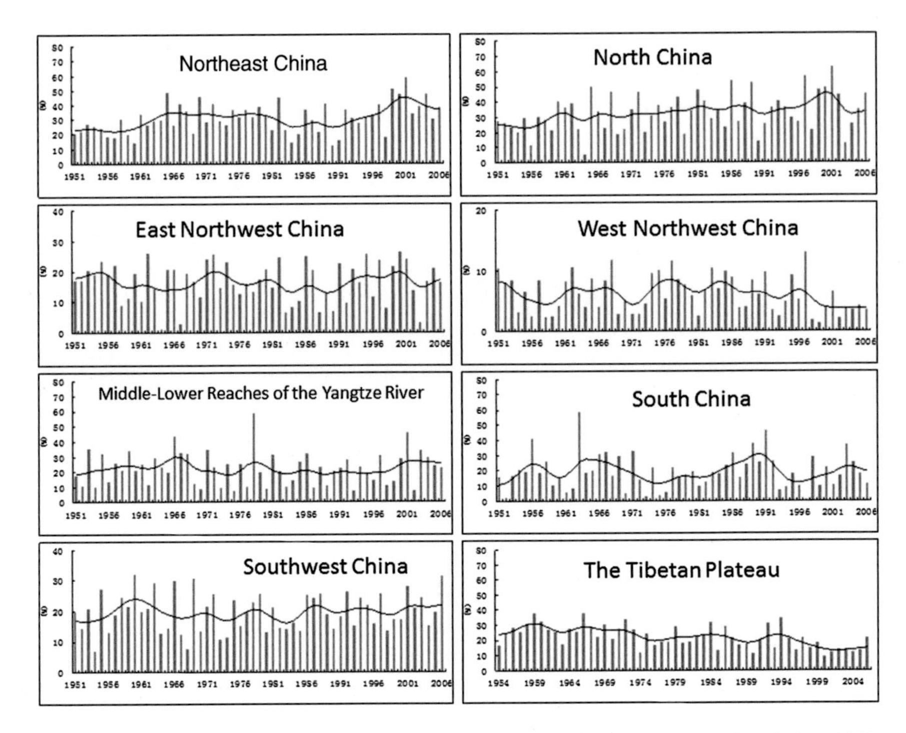

Fig. 2.5 Variations in areal extent (%) of drought over different regions of China during 1951–2006. The smooth curve is generated using an 11-point binomial filter. Because measurements in the Tibetan Plateau began in the early 1950s, the calculation starts from 1954. From Zou et al. (2005)

1951). The ratio of typhoons striking China to the total number of typhoons generated in the northwest Pacific Ocean and South China Sea has increased, especially during the last 10 years, e.g., 50 % in 2010.

2.2.5 Sandstorms

In the past 50 years, the number and severity of severe sandstorms in northern China has decreased slightly (Fig. 2.7). But in individual regions, such as north Qinghai Province, western Xinjiang, and Xilinhot of Inner Mongolia, the number of sandstorms has risen. Wind, precipitation, relative humidity, temperature, and vegetation cover are all important factors that affect the frequency of sandstorms. One of the causes of the decrease in sandstorms in northern China over the past 50 years is the reduction in the averaged near-surface wind speed in those areas and in the number of gale days. In most parts of northern China, sandstorms, precipitation, and relative humidity are negatively correlated. Data show that the amount of summer rainfall in the previous year has an important impact on the occurrence of dust storms.

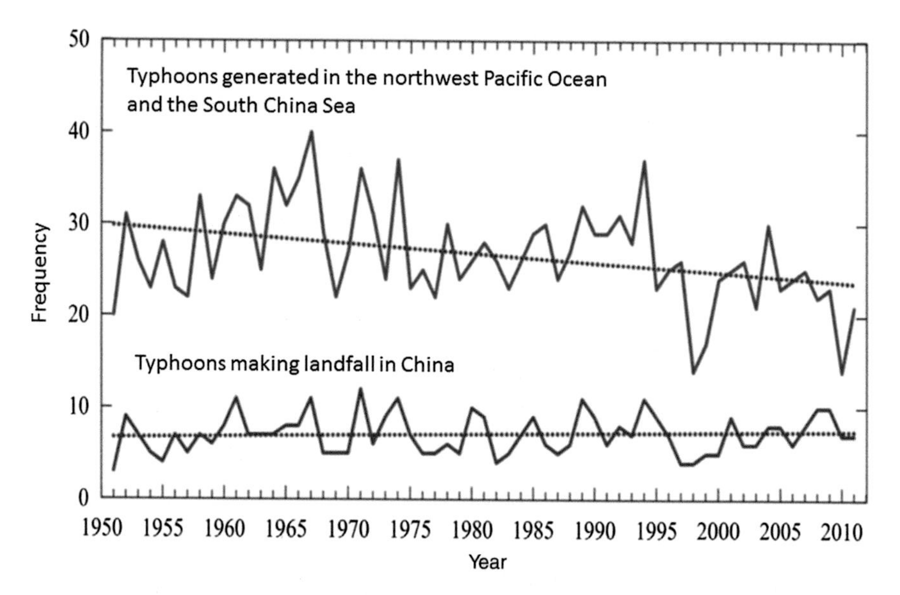

Fig. 2.6 Frequency of typhoons generated in the northwest Pacific Ocean and the South China Sea and making landfall in China during 1951 to 2011 (From China's Climate Change Monitoring Report of 2011, issued by Climate Change Center of China Meteorological Administration 2012)

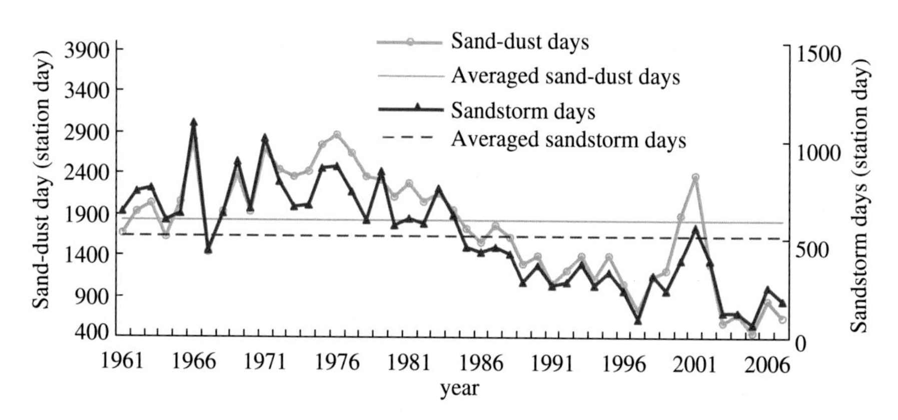

Fig. 2.7 Annual variations in the number of days with sand dust (including blowing sand, sandstorms, and strong sandstorms) and the number of days with sandstorms in springtime in northern China

2.2.6 Other Extreme Weather and Climate Events

2.2.6.1 Hail

China experienced a high incidence of hail events in the 1960s–1980s and significantly fewer events since the 1990s. Across China, hail occurred on average

2.1 days per observational station; it occurred most frequently in 1976 (2.6 days per station) and least frequently in 2000 (1.2 days per station).

2.2.6.2 Wind

Over the past half century, the average number of gale days and the maximum wind speed in China have decreased. Global warming leads to warmer air temperatures in China, which weakens the cold air activities and consequently winds. However, the change in environment of observation stations, e.g., urbanization, during the past several decades, may contribute to the decrease of wind speed on local scale.

2.2.6.3 Fog and Haze

Foggy days have tended to decrease in most areas in China over the past 50 years. Fog events of duration longer than 12 h are mostly concentrated in coastal regions, north China, eastern Gansu and Shanxi, the Sichuan Basin, Yunnan, and Guizhou. However, between 1957 and 2005, haze has increased, and the averaged visibility has decreased by about 10 km in eastern China, at a rate of approximately 0.2 km per year. The magnitude and rate of decline in visibility in western China is half of that in the east, showing that the regional haze problem is getting worse.

2.2.6.4 Thunder and Lightning

In recent decades, the number of days with lightning has decreased across most of China. This decrease is greater in the Tibetan Plateau area and in southern China than in Xinjiang and northern China.

2.2.7 The Relationship Between Global Warming and Extreme Climate Events

In recent decades, the frequency of extreme cold events has been decreasing in China, but the frequency of extreme warm events has been increasing accompanying with global warming. This change of increase in climate fluctuations may be related to global climate change. Climate change can change large-scale global circulation patterns and thereby influence the occurrence frequency of extreme weather and climate events in different regions. Global warming can increase surface evaporation and the saturation vapor pressure, and warmer air contains more water vapor in the atmosphere. It is likely to have more frequent and heavier precipitation at a certain time and place where atmospheric movement induces the convergence of water vapor, leading to more floods. And in other larger areas, due to the increase of surface evaporation, drought becomes more probable (Table 2.1).

Table 2.1 Definitions of extreme weather and climate index and their observed changing trend

Extreme weather and climate index	Definitions	Changing trend
Cold wave	Cool weather in which minimum air temperature dropped by more than 10 °C in 24 h and below 5 °C	The number of cold waves decreased
Frost	Extreme weather that causes plants to damage due to the air minimum temperature dropping below 0 °C	The number of frost days decreased
Heat wave	High temperature (≥35 °C) lasts for over 5 consecutive days	The number of days with daily maximum temperature ≥35 °C slightly increased with significant interdecadal variations
Extreme precipitation	Daily precipitation exceeds 90 % of normal	Extreme precipitation in the Yangtze River valley and to its south increased. The intensity and frequency of extreme precipitation in north China significantly decreased
Continuous rainfall	The wet and cold weather in early spring and late autumn with continuous rainfall and less sunshine over several days or even months	Continuous rainfall days decreased in eastern China and slightly increased in western China
Drought	Little rainfall and dry weather with large surface evaporation which damages crop growing	Drought increased in northern China
Tropical cyclone	Non-frontal vortex in tropical and subtropical oceans, with organized convection and cyclonic circulation, with surface wind speeds ≥10.8 m/s	The number of tropical cyclones generated in the northwest Pacific Ocean and South China Sea decreased, so decreased the number of landfall tropical cyclones on China
Dust storm	An unusual, frequently severe weather condition characterized by strong winds and dust-filled air over an extensive area with the visibility less than 1000 m	The number of sandstorm days decreased
Hail	Weather with precipitation in the form of balls or irregular lumps of ice	Hail days decreased
Gale	Wind speeds greater than 17 m/s	The number of gale days decreased
Fog	Water droplets suspended in the atmosphere in the vicinity of the Earth surface which affect visibility less than 1 km	The number of fog days decreased
Haze	Particles suspended in air, reducing visibility by scattering light less than 10 km	The number of haze days increased
Thunderstorm	Weather with lightning, a transient, high-current electric discharge with path lengths measured in kilometers	The number of thunderstorm days decreased

2.3 Changes in the Cryosphere and the Oceans

Since the 1960s, the glaciers in China have retreated and thinned. The temperature of permafrost has increased and the active layer has thickened. The areas of seasonal permafrost have shrunk, and the thickness has thinned. The extent and amount of snow cover have significantly changed. Sea level is rising along China's coastlines, and sea surface temperature is rising.

2.3.1 Changes in the Cryosphere

Glaciers, permafrost, and snow cover are the main components of China's cryosphere. Sea ice, river ice, and lake ice account for very little proportion. Over half of China's territory is covered by glaciers, permafrost, and/or stable snow cover. The change in cryosphere in China may have a major impact on water resources, the surface water cycle, climate, and engineering projects.

China's cryosphere has been shrinking since the 1960s and 1970s, with 80 % of glaciers in China retreating, at a thinning rate of 0.2–0.7 m/year. For example, according to the long-term observations of Glacier No. 1 at the headwaters of the Urumqi River in the Tianshan Mountains as well as the prediction of a dynamic glacier model, this Glacier No. 1 will disappear in the coming 70–90 years, even disappear within 50 years if the air temperature increases faster than in the past. On the Tibetan Plateau, the ground has been warming and the permafrost active layer has been thickening, with a decrease in the maximum permafrost depth. The mean seasonal permafrost thickness over the Tibetan Plateau has decreased by 10 cm between the 1960s and 1980s and 1980s and 2000s. The maximum permafrost depth has risen by 20–80 m. But the thickening of active layer is quite different across the regions over the Tibetan Plateau. The permafrost temperature has increased by 0.1–0.5 °C along the Qinghai–Tibet Highway from the 1970s to 1990s. This loss of permafrost will lead to a negative influence on ecosystems at future climate scenarios.

Due to the influence of monsoons on China's climate, the change in snow cover across China differs from the Eurasia where snow cover is currently decreasing. There is a trend of decreasing snow cover in the northeast China and increasing snow cover in the northwest China, with significant interannual variations. Snow cover observations are scarce in the plateau regions, and this deficiency together with the error in satellite monitoring shows various snow cover datasets, which bring difficulties about the accurate assessment of the change in snow cover.

Monitoring and research of river ice, lake ice, and sea ice are still relatively weak in China, but the current data show that all of these types of ice are decreasing, in line with global warming.

The latest international research studies on sea-level change and its influencing factors showed that, since 2003, the main reason for global sea-level rise is the melting of cryosphere. An analysis of the total amount of cryospheric melt water in China, from a rough estimate of the areal extent of glaciers, concluded that the potential contribution of the Chinese cryosphere to rising sea level is 0.14 mm/year to ~0.16 mm/year, with a major contribution from glaciers (~0.12 mm/year). In the glacier contribution, glacial melt water supply to river outflows contributed to only 0.07 mm/year to sea-level rise, accounting for about 6.4 % of the global mountain glacier and ice cap contributions to sea-level rise.

2.3.2 Change in the Offshore Environment and Sea Level

The physical and biogeochemical environment alone, the China's offshore has experienced significant changes in recent decades, including the temperature, salinity, circulations, sea level, marine biogeochemical processes, and coastal lines. Sea surface temperature of the offshore of China has been rising, most notably on the continental shelf. The salinity of the Bohai Sea has increased significantly, although salinity changes in other sea areas are not obvious. Offshore winds have weakened, and heat flux and fresh water flux have decreased.

Between 1981 and 2010, the average rate of sea-level rise of offshore China was 2.6 mm/year, higher than the global average by 0.8 mm/year. During that time, average sea-level rise in the Bohai Sea was 2.3 mm/year, 2.6 mm/year in the Yellow Sea, 2.9 mm/year in the East China Sea, and 2.6 mm/year in the South China Sea. Over the next 30 years, the Chinese coastal sea level is expected to rise by 80–130 mm higher than that in 2008 (From People's Republic of China Ministry of Land and Resources, 2010 Sea Level Monitoring Bulletin 2011).

Chinese estuaries are eutrophic, and the Yangtze estuary and adjacent coastal waters have low oxygen levels. In summer, the dissolved oxygen solubility in the hypoxic area of the Yangtze River and adjacent offshore regions is as low as 0.5–1.0 mg/L. Increased CO_2 emissions have also led to acidification of Chinese offshore waters compared to the global median ocean water pH of −0.06. Change in China's coastal waters is, however, also related to human activity, which may exacerbate changes caused by global warming.

2.4 Past Climate Change

Over the past 200–300 million years, global climate has changed in response to continental drift and other environmental factors. Proxy climate records, such as sediment cores, ice cores, tree rings, coral records, and historical documents, have provided information about these past climate changes.

2.4.1 New Findings from the Study of Paleoclimate Proxy Records

The last interglacial–glacial cycle took place 130,000 years ago, and records from Gansu Gulang loess and Shanxi Sanmenxia loess reveal a series of climate change events at the timescale of 1000 years, similar to that in the northern Atlantic Ocean.

Climate events known as the Last Glacial Maximum, and the Younger Dryas events occurred in the past 20,000 years. From the proxy records, such as accumulation of loess, lake sediments, peat deposition, cave deposits, and ice cores, the Younger Dryas event has been identified in China. The $\delta^{18}O$ values from a cave stalagmite in Beijing show that the abrupt climate change in Younger Dryas occurred with almost the same features in northern China as in Greenland region (Fig. 2.8), except for about 80 years later.

The most recent 10,000 years of Earth climate history is called the Holocene megathermal period. The surface air temperature during the early and middle stages of the Holocene was significantly higher than modern times. 7000–6000 years before present was the warmest epoch of the Holocene, with air temperature 1.5 °C higher than present. However, the temperature decreased rapidly 5000 years ago and did not rise again until modern times. Purog Kangri ice cores from the Tibetan Plateau reveal a linear decline of $\delta^{18}O$, indicating a decline in temperature since the Holocene megathermal period. Even during the warm periods, however, there still was occurrence of cold events. Records from a Guliya ice core imply there was a cold event 8200 years ago, with quickly decreasing temperature and a slow subsequent rise; the lowest temperature reached 7.8–10 °C. The warmest phases of the Holocene megathermal period coincided with the flourishing of primitive agriculture in China, when the northern limit of farming and rice production was located north of 2–3° latitudes than the present. At the same time, the inland lakes in northern China, Xinjiang, and Tibet became desalting with high lake level. During the past 2000 years, there have been 100-year fluctuations from cold-warm to dry-wet climate in China. Temperature data have confirmed the occurrence of the Medieval warm period, the Little Ice Age, and climate changes in the twentieth century. The Medieval warm period occurred from A.D. 930 to 1310, with a temperature rise of +0.18 °C (the warmest period was from 1230 to 1250, when temperatures rose 0.9 °C); the Little Ice Age was from A.D. 1320 to 1910, with a temperature decline of −0.39 °C (the coldest period was from 1650 to 1670, when temperature dropped by 1.1 °C). Since A.D. 1920, eastern China has rapidly warmed, the temperature 0.2 °C warmer than the average value over the past 2000 years. There were corresponding cold-warm periods in the area of northwest China and Tibet, but these periods varied from those in eastern China.

In conclusion, paleoclimate data reveal that climate warming in the twentieth century is not as significant as that during the Medieval warm period, but the twentieth century is the warmest period of time in past five hundred years.

Cycles of dry and wet periods over the past 2000 years have varied regionally. Lake sediment records in arid and semiarid areas of China show that the semiarid

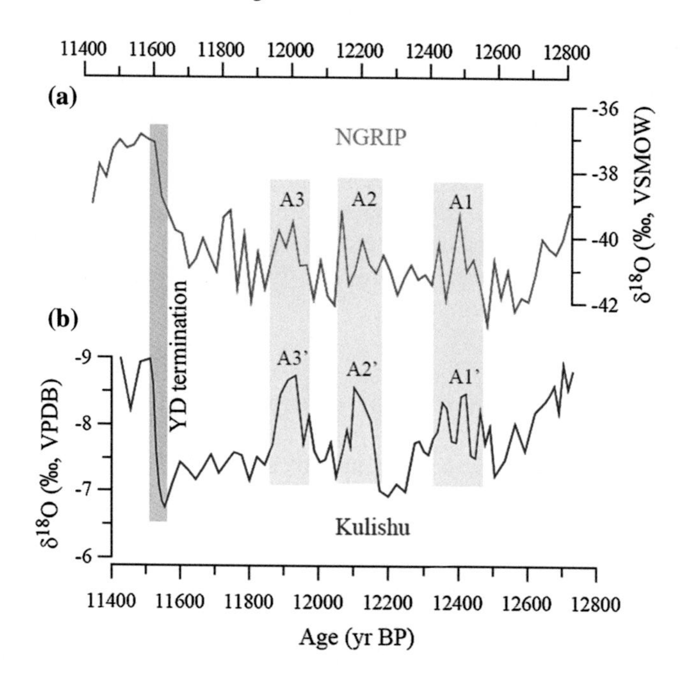

Fig. 2.8 The comparison of oxygen isotope from a stalagmite in Kulishu cave (Beijing, China) with that from an ice core in the North Greenland Ice Core Project (NGRIP). *VPDB* Vienna Peedee belemnite; *VSMOW* Vienna Standard Mean Ocean Water

areas became drier over the past 2000 years, while the arid areas became wet. The centennial-scale changes in climate in semiarid areas were from warm-wet to cold-dry and from cold-wet to warm-dry combination in the arid regions. However, in the eastern humid and semi-humid climate zones, the climate during the Medieval warm period was not persistently droughty; instead, it alternated between drought periods and rainy periods, which lasted several decades. In the most recent 1000 years, a wide range of persistent drought events have occurred in the context of cold climate and their severities were worse than the drought events in the rapidly warming twentieth century.

Over the past 500 years, precipitation was mainly distributed across northern China, occurring more frequently in the warm climates than in cold climates.

2.4.2 Advances in Numerical Simulation of Paleoclimate

2.4.2.1 Simulation of the Last Glacial Maximum and the Holocene warm period

Modeling of Last Glacial Maximum climate using the PMIP2 model (the second phase of Paleoclimate Modeling Intercomparison Project) showed that the weakening of the East Asian summer monsoon is derived from variations of sea surface temperature (SST) and sea ice extent, and changes in the winter monsoon can be attributed to changes in SST, sea ice extent, ice cover on land, and terrain. Increased vegetation is likely to cause a decline in temperature and precipitation in eastern China, and a possible increase of precipitation in western China. Plateau ice cover leads to the cooling of East Asia and weakens the EAM. The distribution pattern of summer precipitation in East Asia in the Holocene is similar to that at the end of the twenty-first century, and this can be used to predict future summer rainfall changes, to some extent.

2.4.2.2 Climate change simulation over the past 1000 years

Modeling has shown that solar activity and volcano eruptions are the main factors causing the Little Ice Age and Medieval warm period. The Little Ice Age (A.D. 1450–1850) and the Medieval warm period (A.D. 1030–1240) occurred during weak and strong periods of global monsoonal precipitation, respectively. NCAR CCSM2 modeling and analysis of summer monsoon precipitation changes show that the root cause of centennial-scale cycles in East Asia precipitation was aperiodic oscillation of solar activity, and interannual and interdecadal variations were likely to be related to feedback within the climate system. The phenomenon of southern floods and northern droughts in recent decades has no climatic precedent.

References

Climate Change Center of China Meteorological Administration. (2012). China's Climate Change Monitoring Report of 2011.

Guo, Q., Cai, J., Shao, X., & Sha, W. (2003). Interdecadal variability of East-Asian summer monsoon and its impact on the climate of China. *Acta Geographica Sinica, 58*, 569–576. (In Chinese with English abstract).

People's Republic of China Ministry of Land and Resources. (2011). 2010 Sea Level Monitoring Bulletin (in Chinese).

Tang, G. L., & Ren, G. Y. (2005). Reanalysis of surface air temperature change of the last 100 years over China. *Climatic and Environmental Research, 10*(4), 791–798. (In Chinese with English abstract).

Wang, S., Gong, D., Ye, J., & Chen, Z. (2000). Seasonal precipitation series of Eastern China since 1880 and the variability. *Acta Geographica Sinica, 55*(3), 281–293. (In Chinese with English abstract).

Wang, S., Ye, J., Gong, D., Zhu, J., & Yao, T. (1998). Construction of mean annual temperature series for the last one hundred years in China. *Quarterly Journal of Applied Meteorology, 9*(4), 392–401. (In Chinese with English abstract).

Wen, X., Wang, S., Zhu, J., & David, V. (2006). An overview of China climate change over the 20th century using UK UEA/CRU high resolution grid data. *Chinese Journal of Atmospheric Sciences, 30*, 894–904. (In Chinese with English abstract).

Zou, X. K., Zhai, P. M., & Zhang, Q. (2005). Variations in droughts over China: 1951–2003. *Geophysical Reseach Letters, 32*(4), L04707. doi:10.1029/2004GL021853.

Chapter 3
The Attribution of Climate Change and Its Uncertainty

Guangyu Shi, Yong Luo, Xiaoye Zhang, Guirui Yu, Renhe Zhang, Xuejie Gao and Wenjie Dong

Abstract Earth's climate has undergone changes caused by both natural and human factors. The natural factors include solar variability, volcanic eruptions, and interactions within the climate system. The human factors affecting climate are the changes in the composition of the atmosphere (including increases in the concentrations of greenhouse gases and aerosols) caused by industrial and social activities, and changes in land use and coverage. For the past century, the effect of solar variability on climate (climatic forcing) has been 1 % less than that of human

G. Shi · X. Gao
Institute of Atmospheric Physics, Chinese Academy of Sciences, P. O. Box 9804
Chaoyang District, Beijing 100029, China
e-mail: shigy@mail.iap.ac.cn

X. Gao
e-mail: gaoxuejie@mail.iap.ac.cn

Y. Luo
Center for Earth System Science, Tsinghua University, Haidaian 100084, Beijing, China
e-mail: Yongluo@mail.tsinghua.edu.cn

X. Zhang (✉) · R. Zhang
Chinese Academy of Meteorological Sciences, Haidian 100081, Beijing, China
e-mail: xiaoye@cams.cma.gov.cn

R. Zhang
e-mail: renhe@cams.cma.gov.cn

G. Yu
Key Laboratory of Ecosystem Network Observation and Modeling,
Institute of Geographic Sciences and Natural Resources Research,
Chinese Academy of Sciences, Beijing 100101, China
e-mail: yugr@igsnrr.ac.cn

W. Dong
Zhuhai Joint Innovative Center of Climate-Environment-Ecosystem, Beijing Normal University, Beijing 100875, China
e-mail: dongwj@bnu.edu.cn

D. Qin et al. (eds.), *Climate and Environmental Change in China: 1951–2012*,
Springer Environmental Science and Engineering,
DOI 10.1007/978-3-662-48482-1_3

activities; the effect of energy within the Earth (geothermal flows) is currently 6 % of that of human factors. On interannual and interdecadal scales, internal climate system factors and their interactions, especially ocean–atmosphere interactions, may be moderately related to global temperature changes; however, it is difficult to determine their contribution to climate change on scales longer than 100 years. Most recent global warming can be essentially attributed to human activities, although uncertainties still exist because of the limitations in observational data, the gaps in our current understanding of the way physical processes affect climate systems, and the various scenarios of future changes.

Keywords Climate change · Natural cause · Anthropogenic driver

3.1 Natural Causes of Climate Change

3.1.1 Solar Variability

Solar variability is related to Earth's climate because the Sun is the major energy source driving the heat engine of Earth's weather–climate system at the mean Sun–Earth distance, the solar energy received by a unit area normal to the solar beam at the top of the atmosphere is called the solar constant (total solar irradiance, TSI). It is the most important physical parameter in meteorological and climatological studies. Modern satellite (or spacecraft-based) climate observations began around 1980, and thus cover only 3 solar periods. To build a complete data base of TSI requires longer observations, and also careful calibration among satellite-borne radiometers. Nevertheless, the basic features of solar radiative energy within the 11-year solar cycle are well known.

To study the radiative forcing of solar variability on longer timescales, the TSI has to be reconstructed with the aid of alternative data. The timescales of currently available surface indicators and historical data are as follows: sunspot numbers (back to 1610), facula index (1950–present), 10.7 cm radio flux (1974–present), and CaIIK 1A index (1976 to present). Among these, the sunspot number has longest time series (almost 400 years).

The satellite data show that the variation of TSI was less than 0.1 % (0.08 %) within an 11-year period during 1978–1990, that is, the ascending phase of the period 21 and the descending phase of the period 22. A reconstruction with sunspot number shows a TSI change of 0.24 % between the Maunder Minimum (A.D. 1645–1715) and the present, or $\Delta F = 1370 \times 0.1\,\% = 1.37\,\mathrm{Wm}^{-2}$ in terms of irradiance for an 11-year period, and $\Delta F = 1370 \times 0.24\,\% = 3.29\,\mathrm{Wm}^{-2}$ for the long term, assuming the solar constant, S_0, is equal to 1370 Wm^{-2}.

These values represent the change in solar radiation reaching the top of the atmosphere. Because the ratio of Earth's surface area to its cross section is 4, and its planetary albedo is ~ 0.3, the values must be multiplied by $(1-\alpha_p)/4 = 0.7/4 = 0.175$ to get the solar radiative forcing. That is to say, within an 11-year period, the solar forcing is $\sim 1.37 \times 0.175 \approx 0.24$ Wm^{-2}; not relevant to the long-term change in Earth's climate. The long-term change of TSI is $\Delta F = 3.29 \times 0.175 \approx 0.6\,Wm^{-2}$. It is noteworthy that this figure accounts for the period since the Maunder Minimum; for the period since A.D. 1850, it should be reduced by about half, to ~ 0.3 Wm^{-2}.

Recent studies show that there has been a 0.05 % increase in TSI over an 11-year smoothed time series between A.D. 1750 and the present, corresponding to an RF of +0.12 W m^{-2}.

In summary, an estimation of solar forcing on climate change since the Industrial Revolution is between +0.3 and +0.12 Wm^{-2} (the former was adopted by IPCC SAR and TAR, and the later by AR4: Houghton et al. 1996; IPCC 2001; Forster et al. 2007). However, this is smaller than the anthropogenic radiative forcing since A.D. 1750, by anywhere from a factor of 5 to an order of magnitude; therefore, TSI is unlikely to contribute significantly to the explanation of global mean temperature rise.

Other possibilities exist, of course. For example, a slight change in extra terrestrial solar irradiance might be amplified by some physical or chemical mechanisms in the Earth-atmosphere system. Therefore, an "amplifier" may be needed to reveal the relationship between solar variability and Earth's climate, including (1) an optical amplifier, (2) a cosmic ray flux–global cloudiness amplifier, (3) an atmospheric electric (or magnetic) amplifier, and (4) a physical–chemical amplifier. However, these amplifiers require further study.

3.1.2 *Volcanic Eruptions*

Volcanic activity is another important process that affects the radiation budget of the Earth-atmosphere system and global climate.

Volcanic eruptions inject particulate matter, such as volcanic ash or debris (mainly composed of lava/basalt), into the atmosphere. The suspension periods of the particles in the atmosphere are from weeks to months at most, which limits their effects on climate.

Among the gases released in volcanic eruptions, H_2O and CO_2 are important greenhouse gases. However, their current concentrations in the atmosphere are so high that no matter how strong an eruption is, its influence will be negligible (it has been estimated that the CO_2 from volcanic eruptions is less than 1 % of the current atmospheric load).

The most significant climate effect of volcanic eruptions is the release of sulfur-containing gases (mostly SO_2 and H_2S) and the subsequent formation of sulfate aerosols. The stratospheric aerosols produced by volcanic eruptions have many effects on atmospheric radiative processes. While they cool the surface by increasing the planetary albedo, to reduce the solar energy reaching Earth's surface, they warm the upper portion of volcanic clouds by absorbing near-infrared solar radiation.

During the past century, there have been only 5–6 prominent volcanic eruptions, followed within 1–2 years by global temperature decreases of 0.1–0.2 °C. The radiative forcing has been estimated as approximately −0.2 Wm^{-2} globally, and −0.3 Wm^{-2} for the Northern Hemisphere (Forster et al. 2007); slightly less than that from anthropogenic aerosols. It has to be noted that the influence of even a very strong volcanic activity on surface temperature can last only a few years; therefore, it is hard to estimate their long-term climate effects.

3.1.3 Geothermal Flow

Geothermal heat can be considered as a natural climate forcing factor. The available data show that geothermal flow is 0.101 and 0.065 Wm^{-2} for oceans and land, respectively, with a global mean of 0.087 Wm^{-2}, which is ~0.06 of anthropogenic climate forcing (Forster et al. 2007).

3.1.4 Internal Factors and Their Interactions

On interannual and interdecadal timescales, internal climate system factors and their interactions, especially the ocean–atmosphere–land interaction, may be linked to global temperature change. El Niño and La Nina events can influence global temperature by 0.1–0.3 °C.

In China, climate change is caused by external forcings as well as interactions among internal climate factors. The interdecadal variations of summer climate over China are influenced to a great extent by the land and ocean processes (Zhang et al. 2013a, b). The significant interdecadal warming of the tropical Indian Ocean and the west Pacific Ocean has caused the interdecadal southward shift of the summertime precipitation belt across China, by influencing the location of subtropical high over the northwest Pacific Ocean (Zhang et al. 2008; Zhou et al. 2009; Han and Zhang 2009). The increasing of the summer precipitation over Yangtze River valley and decreasing over northern China since the middle 1970s are related to both Asian–Pacific Oscillation (APO) (Zhao et al. 2007) and Pacific Decadal Oscillation (PDO) (Yang et al., 2005). The North Pacific Oscillation (NPO) and the wintertime monsoon of China have significant interdecadal variation, resulting in a weakening trend of the wintertime monsoon (Wang et al. 2007). The Atlantic Multidecadal Oscillation (AMO) can cause significantly different distributions of precipitation and temperature across China (Lu et al. 2006). The interdecadal variation of Eurasian snowcover can cause interdecadal variation in the summertime precipitation in northeast and south China (Wu et al. 2009).

3.2 Anthropogenic Causes of Climate Change

3.2.1 Changes in Atmosphere Composition

Earth's atmosphere is continually evolving. Before the Industrial Revolution (~250 years ago), this evolution was mainly dominated by natural causes. Since then, especially in recent decades due to the rapid development of industrial and social activities, the composition of Earth's atmosphere has experienced striking changes. Earth's atmospheric composition is governed by geological, biological, and chemical processes; however, the impact of anthropogenic processes has been increasing over the past few decades. The most prominent example is the increase in atmospheric CO_2, CH_4, other greenhouse gases, and sulfate aerosols. This change may profoundly Earth's climate and environment affecting atmospheric radiation (Table 3.1).

Water vapor is the most important greenhouse gas in the atmosphere, from the viewpoint of impact on Earth's climate, and its content varies greatly across regions and seasons. Since the 1970s, the water vapor content in many parts of the Northern Hemisphere has increased significantly. In current global climate change research, water vapor changes are not treated as a "direct force," but as a "feedback mechanism" (Boucher et al. 2013).

The average global atmospheric CO_2 concentration increased to 386.8 ppm in 2009, with an average annual increase of ~2.0 ppm/year in the past 10 years (1997–2007). Atmospheric halogen-containing compounds such as chlorofluorocarbons (CFCs), hydrogenation of chlorofluorocarbons (HCFCs) and halogenated hydrocarbons, halons, bromine-containing halogenated hydrocarbons have current atmospheric concentrations less than 1 ppb, which is very low compared with

Table 3.1 The major greenhouse gases affected by human activities and their concentration (Solomon et al. 2007)

Greenhouse gas	CO_2	CH_4	N_2O	CFC-11	CFC-12	HCFC-22	CF_4
Before Industrial Revolution (A.D. 1750)	278 ppmv	715 ppbv	270 ppbv	0	0	0	40
Atmospheric content (in 2005)	379 ppmv	1774 ppbv	319 ppbv	251 pptv	538 pptv	169 pptv	74 pptv
Atmospheric content change (1998–2005)	13 ppmv	11 ppbv	5 ppbv	−13 pptv	4 pptv	38 pptv	–
Life of greenhouse gas (year)	50–200[a]	12	114	45	100	12	50,000

[a]Data from IPCC (2001); remaining data from Solomon et al. (2007)

other gases. However, their effects on climate are highly discussed. Because of the signing of the Montreal Protocol in September 1987, to protect the atmospheric ozone layer, and the later Kyoto Protocol, the increase in CFC concentration has slowed, or even decreased. However, as the CFC substitutes, part of the HCFCs and HFCs concentration growth rate was faster, including some species with higher global warming potential cannot be ignored.

The most important recent variations in Earth's atmosphere are a reduction in stratospheric ozone and an increase in tropospheric ozone (Solomon et al. 2007). Annual average concentrations of atmospheric CO_2 across China reached 387.4–405.3 ppm in 2009, with the lowest concentration (387.4 ppm) measured at the Waliguan global base station (36° 17′N, 100° 54′E). This is slightly higher than the global average. These data are broadly consistent with observations in the northern mid-latitudes in the same period (see Fig. 3.1). Concentrations of CH_4, N_2O, and SF_6 also rose in China.

Concentrations of CFC-11, CFC-12, and HCFC-22 are highest in atmospheric halogen-containing gases. Table 3.2 compares the average annual background concentrations at Beijing Shangdianzi station (China) and at five typical background stations in the Northern and Southern Hemispheres in 2007. The difference between the Northern and Southern Hemispheres reflects the various latitudinal emissions from human activities, atmospheric transport, and mixing effects.

Aerosols can change the radiation balance of the Earth-atmosphere system and thus affect climate, by scattering and absorbing solar radiation. Aerosols also have an indirect effect on the microphysical and radiative properties of clouds.

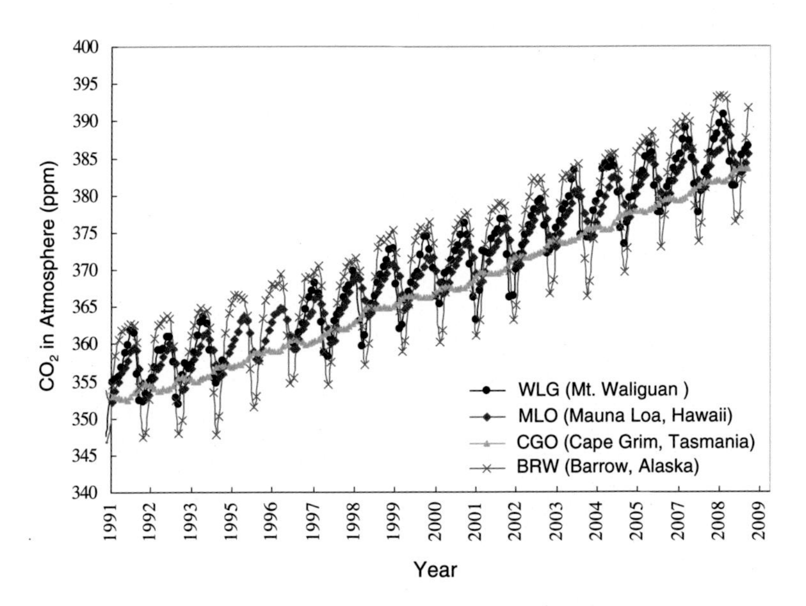

Fig. 3.1 Comparison of changes in atmospheric CO_2 concentration at Waliguan station (China) and three other representative global background stations

Table 3.2 Average annual background concentration of halogen-containing gases

Station	Longitude	Latitude	Height above sea level (m)	CFC-11 (ppt)	CFC-12 (ppt)	HCFC-22 (ppt)
MHD station, Ireland	−9.9	53.33	25	246.29	541.02	194.77
THD station, USA	−124.15	41.05	120	246.44	541.20	195.91
Beijing Shang Dianzi station, China	117.12	40.65	286.5	245.97	541.26	195.35
RPB station, Barbados	−59.43	13.17	45	246.18	540.48	189.13
SMO station, American Samoa	−170.57	−14.24	42	244.86	538.71	176.35
CGO station, Australia	144.68	−40.68	94	243.84	537.94	173.10

Among the atmospheric aerosols found over China, mineral aerosols are the essential component, accounting for ~35 % of PM_{10}. Sulfate and organic carbon aerosols are other two major components (16 and 15 %, respectively) with important scattering effects, black carbon aerosol makes up ~3.5 % of China's aerosols, nitrate is 7 %, and ammonium salt is ~5 % (Zhang et al. 2012a).

Atmospheric aerosol optical depth (AOD) shows a significant increase over China since the 1960s. Figure 3.2 shows the variation of annually averaged AOD retrieved from visibility data with more than 35 years records, which is subdivided into two periods: 1960–1985 and 1986–2005. The slope of the linear trends

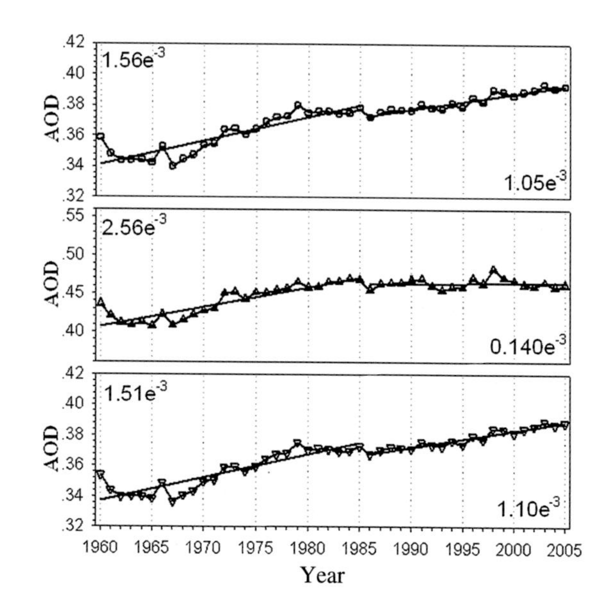

Fig. 3.2 Variations of annual mean AOD over China from 1960 to 2005. The *top* is averaged over 639 stations with data over more than 35 years; the *middle* is averaged over 31 stations located in major cities in China; the *bottom* is averaged over the remaining 609 stations

is marked at the upper left corner and lower right corner of each diagram, respectively.

Before 1985, the atmospheric AOD in China increased significantly, likely tied to the improvement of China's economic development and an increase in the use of coal, oil, and other fossil fuels, especially in metropolitan areas. After 1985, with the promulgation of environmental awareness and more stringent national pollutant discharge standards, this trend has slowed by an order of magnitude, especially in large urban areas.

Using the latest observational data, it can be estimated that the global anthropogenic aerosol direct radiative forcing is -0.23 Wm^{-2} (Zhang et al. 2012b), and the indirect effect is -1.93 Wm^{-2} (Wang et al. 2010). When calculating the influence of black carbon and organic carbon aerosols on cloud, surface temperature, precipitation, and atmospheric circulation in China, as well as the direct climatic effects of the Chinese sulfate and black carbon aerosols and the impacts of the East Asian summer monsoon precipitation, the carbonaceous aerosols are found to cause an increase in the total precipitation in northern China but a reduction in southern China (Zhang et al. 2009), in contrast to the predictions of Menon et al. (2002).

3.2.2 *The Function of the Biogeochemical Cycle*

The CO_2 released to the atmosphere caused by human activity is mainly attributed to fossil fuel combustion and changes in land use since the Industrial Revolution (Ciais et al. 2013). The amount of CO_2 released from fossil fuel combustion increased from 6.4 Gt C (23.5 Gt CO_2)/year in the 1990s to 8.3 Gt C (30.4 Gt CO_2)/year over the period of 2002–2011 (Le Quéré et al. 2013). CO_2 released from land-use changes was estimated to be 1.0 Gt C (3.7 Gt CO_2)/year from 2002 to 2011 (Le Quéré et al. 2013). The 34 Gt CO_2 caused by human activity (assuming the release from land-use change to be steady or slightly decreasing) is reallocated between the atmosphere, the Earth ecosystem, and the ocean carbon pools. This reallocation and exchange between the three carbon pools determine the future atmospheric CO_2 concentration (influenced by global climate change), which in turn affects the Earth's climate system. Therefore, it highly needs to adequately describe the changes in global and regional carbon pools, and to quantitatively evaluate the carbon exchange and the equilibrium relationships between these carbon pools, so as to better understand the mechanism of global climate change and predict the future changes.

Many studies have been carried out on the soil and vegetation carbon reservoirs in China. Table 3.3 shows the carbon density and storage of vegetation and soil, and total carbon storage for different types of ecosystems in China. As shown in the table, the soil and vegetation carbon storage of China is 102.96 and 15.97 Pg C, respectively. The total carbon storage of the above two reservoirs is about 118.93 Pg C.

Table 3.3 Vegetation and soil carbon storage for different types of ecosystems in China

Ecosystem type	Area (10^4 km^2)	Vegetation carbon density (Mg C/hm^2)	Soil carbon density (Mg C/hm^2)	Vegetation carbon storage (Pg C)	Soil carbon storage (Pg C)	Total carbon storage (Pg C)
Forest	142.80	52.28 ± 12.62	156.18 ± 33.29	7.46	22.30	29.76
Grassland	331.00	7.61 ± 4.29	129.92 ± 16.52	2.52	43.00	45.52
Shrub	178.00	18.37 ± 23.86	78.81 ± 48.14	3.27	14.03	17.30
Farmland	108.00	18.48 ± 18.12	103.53 ± 23.77	2.00	11.18	13.18
Desert	128.24	3.71 ± 5.14	61.47 ± 43.52	0.48	7.88	8.36
Wetland	11.00	22.20 ± 5.68	415.10 ± 413.19	0.24	4.57	4.81
Sum	899.04	–	–	15.97	102.96	118.93

Note hm^2 (square of hectometer) = 10^4 m^2; Mg C (megagram of carbon) = 10^6 g C; Pg C (petagram of carbon) = 10^{15} g C. Data sources from Yu et al. (2010)

Since the 1990s, six assessments on the forest vegetation carbon storage have been conducted in China using the national forest inventory data and various statistical methods; however, large differences are found among the six assessments owning to the increase in the forest areas (Fang et al. 2007). China's carbon storage in forest vegetation is approximately 6.24 Pg C, and carbon density is approximately 4.01 kg C/m^2 (Zhang et al. 2013b).

Numerous terrestrial ecosystem carbon cycle studies and regional carbon budget assessments in China indicate that China's terrestrial ecosystem is an important atmospheric carbon sink and plays a crucial role in the regional carbon balance (Yu et al. 2010; Li et al. 2013; Zhu et al. 2014; Fang et al. 2007; Piao et al. 2009). The annual average total carbon sink of forests, grasslands, and shrubs during the period of 1980–2000 is 0.075, 0.007, and 0.014–0.024 Pg C, respectively (Piao et al. 2009). The total amount of carbon sequestrated by ecosystem vegetation and soil during this period is equivalent to 28–37 % of parallel industrial CO_2 emissions in China (Piao et al. 2009). Integrated assessments indicate that the annual atmosphere net CO_2 absorption by Chinese terrestrial ecosystems during the period of 1980–2000 varies from 0.19 to 0.26 Pg C (Piao et al. 2009); the annual average cultivated soil carbon sink is about 16.6–27.8 Tg C (Sun et al. 2010), and the old-growth forest of the northeastern temperate and southern subtropical zones keeps a strong carbon sink function (Yu et al. 2008).

Continuous observations performed by the Chinese Terrestrial Ecosystem Flux Research Network (ChinaFLUX) have been used to determine the typical terrestrial ecosystem carbon sink and source in China (Yu et al. 2013a). The results show that China's alpine meadows and alpine shrub meadows in the Tibetan Plateau have a carbon sink function. Most of the alpine grassland meadows and northern temperate grassland ecosystems are in carbon equilibrium, representing a large interannual fluctuation in net carbon exchange under the influence of precipitation (i.e., weak carbon sink in wet years and a carbon source in dry years). The eastern subtropical natural and planted forests and temperate forest ecosystems have a significant carbon sink function (Yu et al. 2013a, 2014). The strength of carbon sink decreases from south to north due to temperature and water conditions. Net ecosystem productivity (NEP) of the various types of ecosystems is jointly controlled by the mean annual temperature (MAT) and annual total precipitation (MAP). The spatial pattern of NEP across China's terrestrial ecosystems can be evaluated using these two variables (Yu et al. 2013a; Zhu et al. 2014).

China's marine areas are large, but there are fewer studies on the spatial distribution of marine carbon sinks and its overall carbon absorption capacity. Some studies show that China's Bohai Sea, Yellow Sea, and East China Sea are sinks of atmospheric CO_2. Table 3.4 shows ocean-atmosphere carbon exchange rates and strength of marine carbon sinks in China.

It is noteworthy that Earth's climate system has a large thermal inertia because oceans are the large storage of excess heat. Model calculations indicate that Earth currently absorbs as much as 0.85 ± 0.15 W/m^2 more energy from the Sun than that emitted to the outer space. Ocean heat capacity data over the 10 years from 1993 to 2003 confirm this imbalance. This infers that (1) even without any changes in atmospheric composition (such as the complete cessation of anthropogenic emissions of CO_2), it is expected that there will be an additional 0.6° of global warming; (2) the lag of the climate system response to forcing means that surface temperature will take 25–50 years to reach 60 % of balanced response; (3) ice sheet collapse and sea-level rise could accelerate.

Overall, China's existing land and marine data are still not sufficient to fully understand the mechanisms of the regional carbon cycle and accurately assess the regional carbon budgets. Strengthen studies on the land and marine ecosystems carbon cycles and synthetic observations on the land-atmosphere and ocean-atmosphere carbon exchange under the global change in China are still highly needed.

Table 3.4 Marine carbon sinks in China

Sea area	Exchange rate (μmol C/(m^2 s))	Strength of carbon sinks (10^4 t/year)	References
Bohai Sea	0.097	284	Song (2004)
Yellow Sea	0.063	896	Song (2004)
		600–1200	Kim (1999)
East China Sea	0.009	188	Song (2004)
	0.033	726	Zhang et al. (1997)
	0.089	3000 (including the Yellow Sea)	Tsunogai et al. (1999)
		430	Hu and Yang (2001)
South China Sea		1665	Han et al. (1997)

Positive value absorption of atmospheric CO_2 (a sink of atmospheric CO_2). Data sources from Yu et al. (2013b)

3.2.3 Land Use and Land Cover

Changes in land-use and land-cover change (LUCC) will first affect the surface radiation characteristics such as surface albedo. Shortwave albedo varies across different surface covers, such as farmland, water, forests, grasslands, and deserts. And therefore, LUCC will produce climate change radiative forcing by changing the surface radiation balance (IPCC 2001).

Until the middle of the twentieth century, most of the forest areas in the mid-latitudes were decreasing. In recent decades, although Western Europe, North America, and Asia (especially China) have begun to regenerate forests, the destruction of tropical forests has accelerated.

Since the third scientific assessment report by the International Panel on Climate Change (IPCC), a new study has been carried out on the radiative forcing caused by LUCC since the Industrial Revolution. Reconstructed cultivated land-use change data since A.D. 1700 shows that the global radiative forcing caused by cultivated land use is about -0.15 Wm^{-2}. Historical changes in pasture use have also been reconstructed. The total radiative forcing due to changes of cultivated land and pasture since A.D. 1750 is about -0.18 Wm^{-2}.

A notable problem is that changes in surface albedo can influence the radiative forcing of atmospheric aerosols. Therefore, these two aspects need to be considered when estimating the surface albedo effect of LUCC and aerosol radiative forcing.

In addition to the albedo effect, LUCC can also change the surface energy and material exchange, such as the moisture budget, emission of CO_2 and methane, and biomass-burning aerosols and dust aerosols, which in turn disturb precipitation, temperature, and atmospheric circulation, and affect regional and global climate. Global air temperature in response to LUCC depends on the increase in surface albedo in winter and spring, and on the decrease in evaporation in the tropics in summer (increasing the temperature effect). Current estimates of global temperature response caused by past reductions of forests range between 0.01° and 0.25°. Although many studies have shown that LUCC has reduced global temperature over the past 300 years, the rapid destruction of tropical rainforests in recent years has caused temperature increases due to decreased evaporation, and this may be increasingly important to global climate. There are limited data on the climate impacts of LUCC in China, compared with other international regions. Chinese scientists began studying the effects on climate of LUCC, using numerical simulations, in the 1990s. Results indicate that the impacts of large regional vegetation changes on regional precipitation and temperature are significant, and the impacts on temperature are more significant than on precipitation. Vegetation degradation induces local temperature rise, dries the middle and lower layer of the atmosphere, and causes increased near-surface wind speed. Reforestation induces warm winters and cold summers, moist atmosphere, and decreased near-surface wind speed, and reduces dust storms to a certain degree. Regional climate simulations of LUCC in the recent history of China, using a regional climate model, show that since A.D. 1700 LUCC may have had a significant impact on the regional precipitation and temperature over China, including an increase in average temperature in most regions of China after 1900, in particular over the past 50 years. The East Asia winter and summer monsoon circulation has intensified, causing an increase in annual average precipitation in the south and a decrease in the north, reducing significantly the annual average temperature in the south. Simulation results over the past 1000 years using an intermediate complex earth system model show that, when only considering the albedo effect, LUCC in the most recent 300 years induces a global cooling of 0.09°–0.16° and a decrease of 0.14°–0.22° in the annual average temperature in the Northern Hemisphere.

In addition to numerical simulations, direct observational data can be used to study the sensitivity of surface air temperature change on LUCC in China, providing observational support for the numerical simulations. Observations show that the undeveloped areas of China, including deserts, and the areas with active human activities (e.g., east China) have large surface temperature increases, while areas with good vegetation cover have relatively small temperature increases. The climate changes in northwest China in the 1970s and 1990s resulted from changes in vegetation cover. The data show that the temperature and precipitation in these areas and periods, and the related interdecadal variations of atmospheric circulation, have probably been affected by changes in the local vegetation cover. However, any observed characterization of climate change is determined by a variety of factors; therefore, the quantitative impact of LUCC should be further studied.

3.2.4 Radiative Forcing

Radiative forcing acts as the driving force to climate change. Radiative forcing is not only due to the effects of human activity (e.g., increases in atmospheric greenhouse gases and aerosol concentrations, and changes in land use and land cover), but also from natural factors such as volcanic activity and solar radiation cycles.

Total radiative forcing of long-lived greenhouse gases (including CO_2, CH_4, N_2O, and halogenated hydrocarbons) since A.D. 1750 is +2.63 [±0.26] Wm^{-2}; radiative forcing of the stratospheric O_3, tropospheric O_3, and stratospheric water vapor generated by CH_4 is −0.05 [±0.10], +0.35 [−0.1, +0.3], and +0.07 [±0.05] Wm^{-2}, respectively. Total direct radiative forcing and indirect radiative forcing (including only the cloud albedo effect) of atmospheric aerosols (including sulfate, nitrate, biomass burning, organic carbon, black carbon, and mineral dust) are −0.50 [±0.40] and −0.70 [−1.1, +0.4] Wm^{-2}, respectively. Radiative forcing caused by changes in surface albedo resulting from land use and deposition of black carbon aerosols on snow are −0.20 [±0.20] and +0.10 [±0.10] Wm^{-2}, respectively. The cirrus cloud effects induced by aircraft wakes is 0.01 [−0.007, +0.02] Wm^{-2}. Total anthropogenic forcing is +1.6 [−1.0, +0.8] Wm^{-2}.

3.3 Causes of Climate Change

3.3.1 The Uncertainty of Observational Data

Observational instruments were applied in climate observation, and the direct monitoring of environmental elements has made a remarkable progress since the end of the nineteenth century, especially in the observation objective and observational tools, the spatial coverage, multiple source monitoring, and quality control. These observational data were applied in the research of climate system and climate change to a great extent. For example, the trend that the surface air temperature in China has generally risen since twentieth century was acknowledged in the scientific community.

The uncertainties of observational data come from many aspects, such as the limitation in time and space, the change of observational environment, the replacement of observational instruments, and statistical means. For example, the observational data in China during the first half of twentieth century is scarce, while the data quality was influenced by the urbanization during the latest 50 years. The total number of observational stations with more than 100 years is only 71 in China, while all these stations are located in eastern China. The meteorological observation network with nationwide coverage was set up after 1951. At present, there are about 2400 observational stations with systematic and continuous data for more than 60 years, however few of them located in western China. That is the reason why it is very difficult to obtain the standard serials of air temperature in China or on regional scale (Tang et al. 2009).

The process of urbanization can influence the regional climatological observation in several ways: firstly, the urbanization has changed the underlying surface, which can affect the local and regional climate directly; secondly, for the observational stations which do not relocate during the urbanization, their environment around the stations has changed greatly, which makes the observational data are not representative for local climate; thirdly, large amount of greenhouse gases and air pollutants were emitted in urbanization process, which can influent the local climate as well. Research results show that the climate effect by urbanization is still aggrandizing, especially in eastern China.

The recent researches began to concentrate on the effect of the relocation of observational stations and the replacement of instruments. The relocation of observational stations can have a significant impact on data uniformity, especially for the meteorological variables such as wind speed, extreme temperature, and rainfall. Most of the meteorological stations in China have relocated, i.e., about 80 percentage of the ground base stations have moved to other place since 1949, while 70 % of them have moved more than one time. Observational data uniformity can also be affected by the instrument replacement and update and the observation method. The ground meteorological observation standard has been modified a lot since 1949. In the observation of meteorological variables such as temperature, humidity, wind speed, and rainfall, the observational specification has changed largely, as well as the observational instruments, the installation height, and observation time.

3.3.2 Model Data

The analysis of physical causes of climate change based on the driving force (radiative forcing) provides the causes of climate change, while numerical model simulation provides the "result." Analysis of the relationships between the "cause" and "result" will help reveal the actual cause of global climate change.

Based on the well-known physical principles, climate models are able to simulate the contemporary Earth's climate and to reproduce the main features of past climate and climate change and therefore are the primary tool for climate change projections.

Earth's climate is an extremely complex, open, giant system, involving different temporal and spatial scales of interaction, and therefore, it requires multidisciplinary integration of the atmospheric sciences, oceanography, geophysics, chemistry, ecology, mathematics, and computer sciences.

The most uncertain processes in present climate models are as follows:

(1) Parameterization of cloud cover. Although clouds play an important role in Earth's climate, past and future cloud parameters (cloud cover, cloud altitude, optical depth, etc.) are little known and hard to determine; therefore, parameterization schemes in climate models contain a great deal of subjectivity.
(2) Aerosol–cloud–radiation interaction, especially where atmospheric aerosols act as cloud condensation nuclei, affecting droplet size, cloud reflectivity, cloud lifetime, and precipitation (an indirect climatic effect of aerosols). There is a lack of research into these mechanisms and a lack of observational data, seriously affecting existing climate models.
(3) Response and feedback of ecosystems to climate change. This requires the simulation of water vapor and energy exchange, vegetation VOC emissions, and the simulation of sources and sinks of carbon and nitrogen in ecosystems.
(4) The internal variability of the climate system.

Scenarios of possible future greenhouse gas emissions depend on estimates of human factors, such as population, energy consumption, and infrastructure development, policy factors, progress in greenhouse gas emission reduction and removal, and new energy development (e.g., the IPCC SRES scenarios and RCP scenarios.

The uncertainty in estimations of climate change is even greater at the regional and local scale, due to a lack of observational data, and low model resolution.

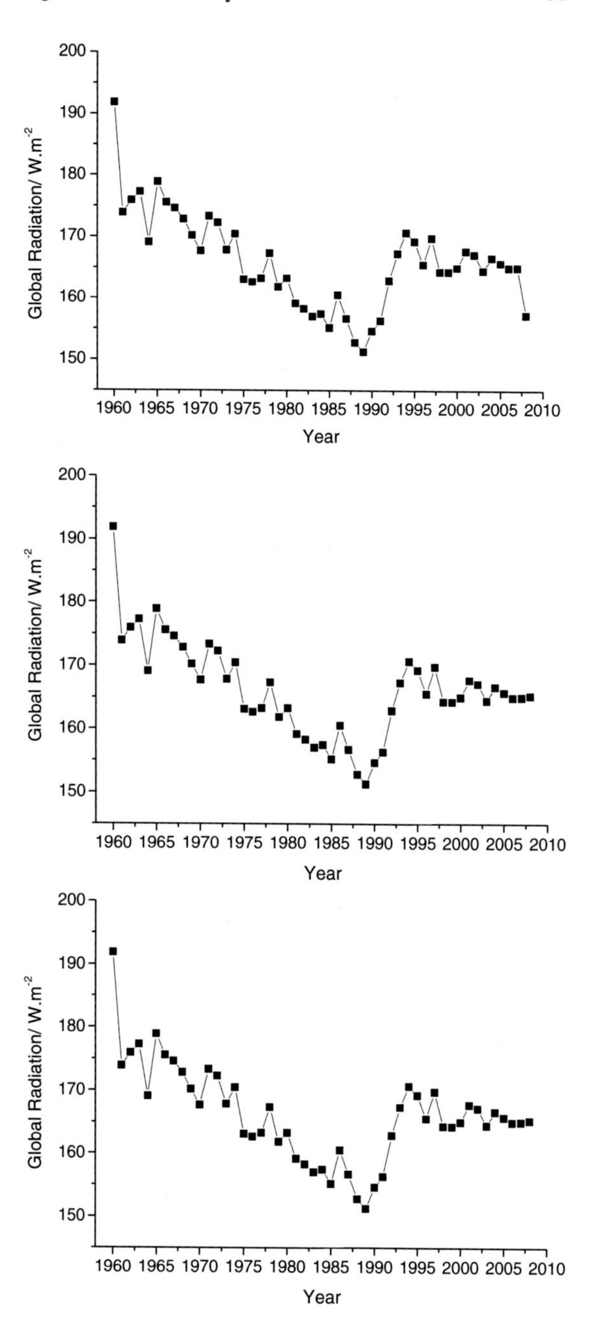

Fig. 3.3 Changes in the total surface solar radiation in China in 1960–2008

3.3.3 *Complexity in Causes of Climate Change*

As noted here, the studies of causes of climate change, regardless of the observational data or model, have shown great uncertainty. The observational data are the basis of the model study, and also the final validation of the model results. However, among the many elements influencing climate, there exist some contradictory in the observational data. Here, we will use the solar radiation, clouds, aerosols, evaporation, etc., in China as examples to explain the difficulty and complexity in attributing climate change to various factors.

The amount of solar radiation reaching the ground (SSR) is an important factor in the surface radiation balance and thus the Earth's climate. Studies have found that regardless of sky conditions and cloud conditions, the global trend of SSR before and after A.D. 1990 began increasing, i.e., the so-called global dimming (dimming) into a global light (brightening). Its climate implication is that before 1990, "dimming" may have had some kind of role in global warming, and the subsequent "brightening" may have accelerated a warming trend. Figure 3.3 shows

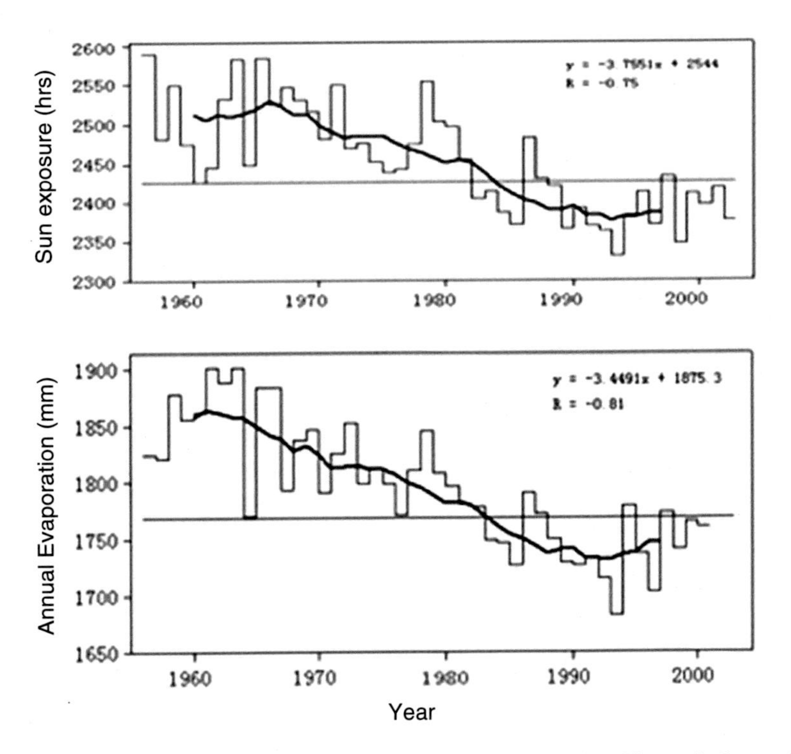

Fig. 3.4 Changes in hours of sunshine (solar radiation) and evaporation. The *top* is the trend of the mean solar hours over China from 1956 to 2002. The *bottom* shows the trend of mean evaporation over China from 1956 to 2002

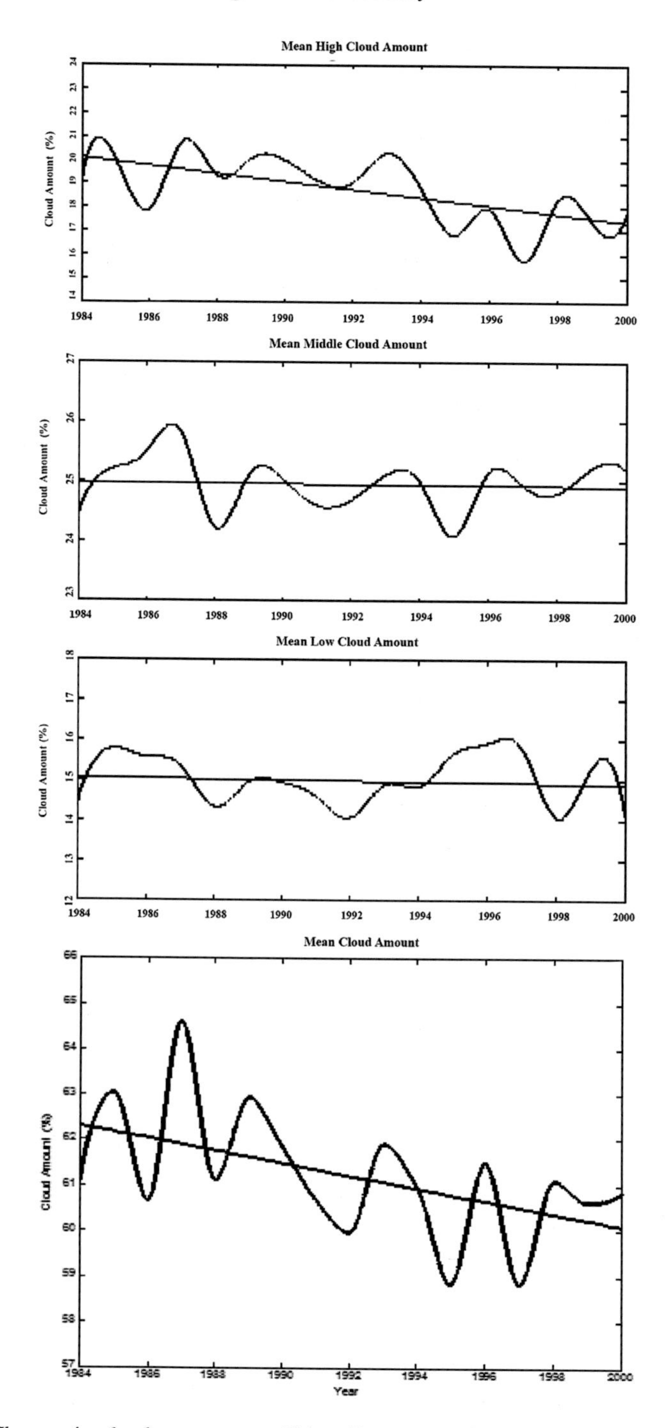

Fig. 3.5 Changes in cloud amount over China. From *top* to *bottom* are high-altitude clouds, medium-altitude clouds, low-altitude clouds, and total clouds

changes in SSR in 1960–2008. It can be clearly seen that SSR increased significantly from 1990, but after 1995, it returned to a decreasing trend.

Interestingly, in roughly the same period (1956–2002), the regional average annual hours of sunshine and evaporation in China showed nearly the same trend (Fig. 3.4). This "synchronous change" is easy to understand, but these changes do not hold for other climate elements.

The main factors affecting SSR are clouds and atmospheric aerosols. Changes in cloud cover (measured by ISCCP satellite data over China from 1984 to 2000) are shown in Fig. 3.5 (the data from meteorological stations show similar results). The reduction in the total cloud-cover results in an increase of solar radiation reaching the ground. The variation of the total cover of medium-altitude clouds and low clouds is not obvious; therefore, the reduction of the total cloud cover is mainly due to the reduction of high-altitude clouds, with a reduction in the range of 2.7 %, accounting for 14.4 % of the mean high cloud cover. It is generally believed that the "greenhouse effect" of high clouds exceeds its "albedo" effect, while cloud-cover reduction will produce a warming effect on the ground. However, on the other hand, the annual average increase of cloud water path is ~ 18 g/m^2 in China, an increase of more than 25 %. Accordingly, the average cloud optical thickness has a significant increasing trend, with an increase of about 1.0. This increase will reduce the SSR and produce a "cooling" effect on the ground.

An increase in atmospheric aerosols (optical thickness) (Fig. 3.2) will lead to a reduction of the SSR, but aerosol variation itself did not create the turning point after 1990.

Another paradox comes from the increase of surface temperature and the reduction of evaporation. The possible cause of this is the weakening of the wind speed in recent decades.

3.3.4 Conclusions

At present, the focus of the controversy on global change is not the higher temperature of Earth's surface over the past 100 years, or the increasing concentration of CO_2, but the interactions between them. In other words, the focus of the controversy lies in the causes of global climate change.

The attribution problem is a complicated issue, relating to both natural and human factors. Changes in solar radiation and geothermy are considered here as factors in global climate change. The climate forcing of solar radiation and geothermy is +0.12 and +0.16 Wm^{-2}, respectively, much smaller than the total climate forcing by anthropogenic factors of +1.6 Wm^{-2}. Climate forcing by volcanic eruptions is about −0.2 Wm^{-2}, and the influence of eruptions only lasts a few years, cooling the Earth for a brief period.

All in all, physical observations and climate models indicate that natural processes and variations in climate cannot explain the climate change over the last 50 years of the twentieth century. Anthropogenic factors are very likely the main cause of global warming.

References

Boucher, O., Randall, D., Artaxo, P., Bretherton, C., Feingold, G., Forster, P., et al. (2013). Clouds and aerosols. In T. F. Stocker, D. Qin, G.-K. Plattner, M. Tignor, S. K. Allen, J. Boschung, A. Nauels, Y. Xia, V. Bex & P. M. Midgley (Eds.), *Climate change 2013: The physical science basis*. Contribution of Working Group I to the Fifth Assessment Report of the Intergovernmental Panel on Climate Change (pp. 571–658). Cambridge: Cambridge University Press. doi:10.1017/CBO9781107415324.016

Ciais, P., Sabine, C., Bala, G., Bopp, L., Brovkin, V., Canadell, J., et al. (2013). Carbon and other biogeochemical cycles. In T. F. Stocker, D. Qin, G.-K. Plattner, M. Tignor, S. K. Allen, J. Boschung, A. Nauels, Y. Xia, V. Bex & P. M. Midgley (Eds.), *Climate change 2013: The physical science basis*. Contribution of Working Group I to the Fifth Assessment Report of the Intergovernmental Panel on Climate Change (pp. 465–570). Cambridge: Cambridge University Press.

Fang, J. Y., Guo, Z. D., Piao, S. L., & Chen, A. P. (2007). Terrestrial vegetation carbon sinks in China, 1981–2000. *Science in China Series D-Earth Sciences, 50*(9), 1341–1350.

Forster, P., Ramaswamy, V., Artaxo, P., Berntsen, T., Betts, R., Fahey, D. W., et al. (2007). Changes in atmospheric constituents and in radiative forcing. In S. Solomon, D. Qin, M. Manning, Z. Chen, M. Marquis, K. B. Averyt, M. Tignor & H. L. Miller (Eds.), *Climate change 2007: The physical science basis*. Contribution of Working Group I to the Fourth Assessment Report of the Intergovernmental Panel on Climate Change. Cambridge: Cambridge University Press.

Han, W. Y., Lin, H. Y., Cai, Y. Y. (1997). Researches on carbon fluxes in south sea. *Acta Oceanologica Sinica, 19*(1), 50–54 (in Chinese).

Han, J., & Zhang, R. (2009). The dipole mode of the summer rainfall over east China during 1958–2001. *Advances in Atmospheric Sciences, 26*, 727–735.

Houghton, J. J., Filho, L. G. M., et al. (Eds.). (1996). *Climate change 1995: The science of climate change*. Cambridge, UK: Cambridge University Press.

Hu, D. X., & Yang, Z. S. (2001). *Key processes of ocean flux in east sea*. Beijing: China Ocean Press. (in Chinese).

Intergovernmental Panel on Climate Change (IPCC). (2001). *Climate change 2001: The scientific basis*. In J. T. Houghton et al. (Eds.), Contribution of Working Group I to the Third Assessment Report of the Intergovernmental Panel on Climate Change (881 pp). New York: Cambridge University Press.

Kim, K. R. (1999). Air-sea exchange of the CO_2 in the Yellow Sea. In *The 2nd Korea-China Symposium on the Yellow Sea Research, Seoul*.

Le Quéré, C., Andres, R. J., Boden, T., Conway, T., Houghton, R. A., House, J. I., et al. (2013). The global carbon budget 1959–2011. *Earth System Science Data, 5*(1), 165–185.

Li, X., Liang, S., Yu, G., Yuan, W., Cheng, X., Xia, J., et al. (2013). Estimation of gross primary production over the terrestrial ecosystems in China. *Ecological Modelling, 261–262*, 80–92.

Lu, R., Dong, B., & Ding, H. (2006). Impact of the Atlantic multi-decadal oscillation on the Asian summer monsoon. *Geophysical Reseach Letters, 33*, L24701.

Menon, S., Hansen, J., Nazarenko, L., & Luo, Y. F. (2002). Climate effects of black carbon aerosols in China and India. *Science, 297*, 2250–2253.

Piao, S., Fang, J., Ciais, P., Peylin, P., Huang, Y., Sitch, S., et al. (2009). The carbon balance of terrestrial ecosystems in China. *Nature, 458*(7241), 1009–1013.

Solomon, S., Qin, D., Manning, M., et al. (2007). *IPCC, 2007: Climate change 2007: The physical science basis [J]*. Contribution of Working Group I to the Fourth Assessment Report of the Intergovernmental Panel on Climate Change, 2007.

Song, J. M. (2004). *Biogeochemistry of China Seas*. Jinan: Shandong Science and Technology Press. (in Chinese).

Sun, W., Huang, Y., Zhang, W., Yu, Y. (2010). Carbon sequestration and its potential in agricultural soils of China. *Global Biogeochemical Cycles, 24*(3), GB3001.

Tang, G., Ding, Y., Wang, S., et al. (2009). Comparative analysis of the time series of surface air temperature over China for the last 100 years. *Advanced Climate Change Research, 5*(2), 71–78. (in Chinese with English abstract).

Tsunogai, S., Watanabe, S., & Sato, T. (1999). Is there a "continental shelf pump" for the absorption of atmospheric CO_2? *Tellus B, 51*(3), 701–712.

Wang, L., Chen, W., & Huang, R. (2007). Changes in the variability of north Pacific oscillation around 1975/1976 and its relationship with East Asian winter climate. *Journal Geophysical Research, 112*, D11110.

Wang, Z. L., Zhang, H., Shen, X. S., Gong, S. L., & Zhang, X. Y. (2010). Modeling study of aerosol indirect effects on global climate with an AGCM. *Advances in Atmospheric Sciences, 27*(5), 1064–1076. doi:10.1007/s00376-010-9120-5.

Wu, B., Yang, K., & Zhang, R. (2009). Eurasian snow cover variability and its association with summer rainfall in China. *Advances in Atmospheric Sciences, 26*, 31–44.

Yang, X., Xie, Q., Zhu, Y., Sun, X., & Guo, Y. (2005). Decadal-to-interdecadal variability of precipitation in North China and associated atmospheric and oceanic anomaly patterns. *Chinese Journal of Geophysics, 48*, 789–797. (in Chinese).

Yu, G., Chen, Z., Piao, S., Peng, C., Ciais, P., Wang, Q., et al. (2014). High carbon dioxide uptake by subtropical forest ecosystems in the East Asian monsoon region. *Proceeding of the National Academy of Sciences of the United States of America, 111*(13), 4910–4915.

Yu, G. R., He, N. P., & Wang, Q. F. (2013a). *Carbon budget and carbon sink of ecosystems in China: Theoretical basis and comprehensive assessment*. Beijing: Science Press. (in Chinese).

Yu, G. R., Li, X. R., Wang, Q. F., & Li, S. (2010). Carbon storage and its spatial pattern of terrestrial ecosystem in China. *Journal of Resources and Ecology, 1*(2), 97–109.

Yu, G. R., Zhang, L. M., Sun, X. M., Fu, Y. L., Wen, X. F., Wang, Q. F., et al. (2008). Environmental controls over carbon exchange of three forest ecosystems in eastern China. *Global Change Biology, 14*(11), 2555–2571.

Yu, G.-R., Zhu, X.-J., Fu, Y.-L., He, H.-L., Wang, Q.-F., Wen, X.-F., et al. (2013b). Spatial patterns and climate drivers of carbon fluxes in terrestrial ecosystems of China. *Global Change Biology, 19*(3), 798–810.

Zhang, Y. H., Huang, Z. Q., Ma, L. M., Qiao, R., & Zhang, B. (1997) Carbon dioxide in surface water and its flux in east China sea. *Journal of Oceanography in Taiwan Strait, 16*(1), 37–42 (in Chinese).

Zhang, C., Ju, W., Chen, J., Zan, M., Li, D., Zhou, Y., et al. (2013a). China's forest biomass carbon sink based on seven inventories from 1973 to 2008. *Climatic Change, 118*(3–4), 933–948.

Zhang, H., Wang, Z., Guo, P., & Wang, Z. (2009). A modeling study of the effects of direct radiative forcing due to carbonaceous aerosol on the climate in East Asia. *Advances in Atmospheric Sciences, 26*, 57–66.

Zhang, X. Y., Wang, Y. Q., Niu, T., Zhang, X. C., Gong, S. L., Zhang, Y. M., et al. (2012a). Atmospheric aerosol compositions in China: Spatial/temporal variability, chemical signature, regional haze distribution and comparisons with global aerosols. *Atmospheric Chemistry and Physics, 11*, 26571–26615.

Zhang, H., Wang, Z. L., Wang, Z. Z., Liu, Q. X., Gong, S. L., Zhang, X. Y., et al. (2012b). Simulation of direct radiative forcing of aerosols and their effects on east asian climate using an interactive AGCM-aerosol coupled system. *Climate Dynamics, 38*, 1675–1693.

Zhang, R., Wu, B., Han, J., & Zuo, Z. (2013b). Effects on summer monsoon and rainfall change over China due to Eurasian snow cover and ocean thermal conditions. In B. R. Singh (Ed.), *Climate change—Realities, impacts over ice cap, sea level and risks* (pp. 227–250). Rijeka: InTech.

Zhang, R., Wu, B., Zhao, P., & Han, J. (2008). The decadal shift of the summer climate in the late 1980s over eastern China and its possible causes. *Acta Meteor Sinica, 22*, 435–445.

Zhao, P., Zhu, Y., & Zhang, R. (2007). An Asian-Pacific teleconnection in summer tropospheric temperature and associated Asian climate variability. *Climate Dynamics, 29*, 293–303.

Zhou, T., Yu, R., Zhang, J., Drange, H., Cassou, C., Deser, C., et al. (2009). Why the western Pacific subtropical high has extended westward since the late 1970s. *Journal of Climate, 22*, 2199–2215.

Zhu, X.-J., Yu, G.-R., He, H.-L., Wang, Q.-F., Chen, Z., Gao, Y.-N., et al. (2014). Geographical statistical assessments of carbon fluxes in terrestrial ecosystems of China: Results from upscaling network observations. *Global and Planetary Change, 118*, 52–61.

Chapter 4
Impacts of Climate Change on the Environment, Economy, and Society of China

Yongjian Ding, Mu Mu, Jianyun Zhang, Tong Jiang, Tingjun Zhang, Chunyi Wang, Lixin Wu, Baisheng Ye, Manzhu Bao and Shiqiang Zhang

Abstract This chapter evaluates the characteristics and extent of impacts of modern climate change on the hydrology, ecology, agriculture, health, economy, and society of China. The impacts of climate change on water resources, hydrological processes, the cryosphere, and ocean hydrological processes are analyzed, as well as the impacts on land ecosystems, desertification, and soil erosion. The impact of global sea-level change on marine ecology and the coastal environment is

---

Y. Ding (✉) · B. Ye · S. Zhang (✉)
State Key Laboratory of Cryospheric Science, Cold and Arid Regions Environmental and Engineering Research Institute, Chinese Academy of Sciences, Lanzhou 730000, China
e-mail: dyj@lzb.ac.cn

S. Zhang
e-mail: zhangsq@lzb.ac.cn

M. Mu
Institute of Oceanology, Chinese Academy of Sciences, Qingdao 266071, China

J. Zhang
State Key Laboratory of Hydrology-Water Resources and Hydraulic Engineering, Nanjing Hydraulic Research Institute, Nanjing 210029, China

T. Jiang
National Climate Center, Chinese Meteorological Administration, Beijing 100081, China

T. Zhang
College of Earth Environmental Sciences, Lanzhou University, Lanzhou 730000, China

C. Wang
Chinese Academy of Meteorological Sciences, Chinese Meteorological Administration, Beijing 100081, China

L. Wu
Physical Oceanography laboratory, Ocean University of China, Qingdao 266003, China

M. Bao
Key Laboratory of Horticultural Plant Biology of Ministry of Education, Huazhong Agricultural University, Wuhan 430070, China

D. Qin et al. (eds.), *Climate and Environmental Change in China: 1951–2012*, Springer Environmental Science and Engineering,
DOI 10.1007/978-3-662-48482-1_4

comprehensively assessed. The chapter also summarized the impact of climate change on farming, animal husbandry, forestry, aquaculture, and fisheries.

Keywords Impact · Climate change · Environment · Economy · Society

4.1 Hydrology and Water Resources

The measured discharge of large rivers in China has been decreasing in the past 50 years, with the discharge from the Haihe River and Yellow River being significantly reduced. Human economic activity is still the main reason for the recent changes in river runoff. However, the impact of climate change on discharge is increasing. Along with socioeconomic development, climate change has exacerbated the conflict between water supply and demand in the northern arid regions of China and exacerbated water depletion (due to water quality) of the southern humid regions. Over the last decade, the river discharge in the northwest alpine mountain areas has significantly increased, mainly due to accelerated glacier melting. Because extreme precipitation events have increased, the frequency and volumes of floods have increased, and the scope and intensity of drought have also increased. And sea-level rise has posed serious challenges to flood control in coastal areas.

4.1.1 River Discharge and Water Resources

Hydrological observation records from the Yangtze, Yellow, Songhua, and Zhujiang rivers, spanning nearly 100 years, suggest that discharges of these major rivers are decreasing, at rates from 0.5 % per 10 years to 4 % per 10 years (Fig. 4.1). The measured discharges of the Haihe, Yellow, and Liaohe rivers have also significantly decreased, with the Haihe River discharge reduced ~30–70 % (Zhang et al. 2007). The annual average discharge of the middle and lower reaches of the Yangtze River, the upper reaches of the Huaihe River and Nen River, and rivers in Xinjiang has increased by 2–9 % since the 1980s.

Change in measured river discharge is the product of several environmental factors, including land cover changes caused by human social and economic activities (e.g., construction of water conservation projects and urbanization), climate change, and economic and social development. Overall, human social and economic activities have more significant impacts on river discharge in the northern arid regions of China and have limited impacts in the humid southern areas. For example, human social and economic activities are the major reasons for the decreased discharge in the middle reaches of the Yellow River since 1970. Climatic

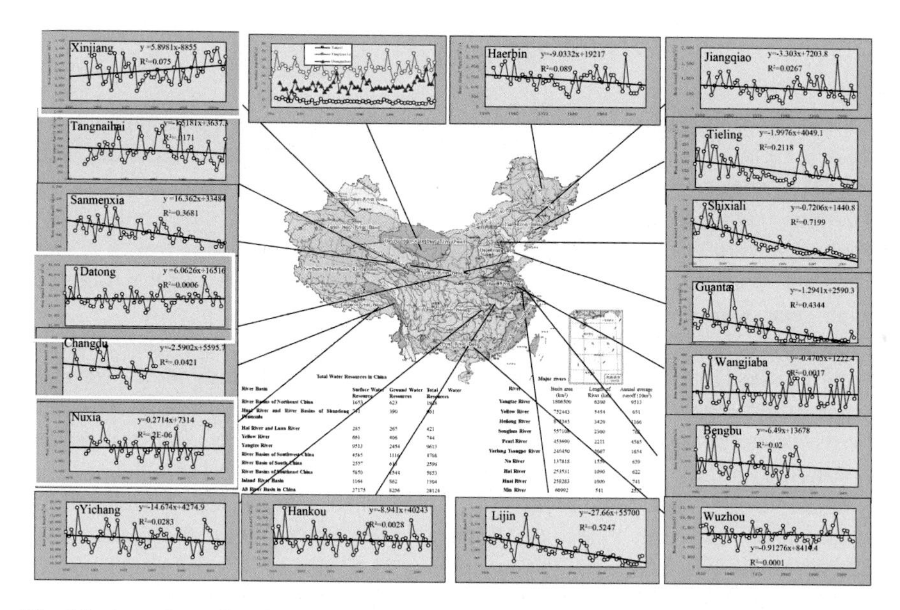

Fig. 4.1 Annual runoff variations of large rivers since 1950 in China (modified from Zhang et al. 2007; Su et al. 2007; Ding et al. 2007)

factors have increasingly contributed to reduced river discharge since the 1980s. Climate change and human social and economic activities accounted for ~30 and 70 % reduction in river discharge, respectively. Climate change exacerbated at some extent the contradiction between water supply and demand in the arid regions of northern China.

The water quality in China is declining. National Water Quality Monitoring and Evaluation results in 2008 showed that the water quality of 45 % of the studied areas met or exceeded the surface water III standard, and 28 % of the areas had serious water pollution, worse than the V standard. River segments with water quality worse than the V standard accounted for 21 % of the total river segments, an increase of 4 % compared with 2001. The rapid development of regional industrial and agricultural activities contributed to the deterioration of water quality, but climate change also had an effect on hydrology. Higher air temperatures effected the biological environment of rivers and lakes, and the distribution of water organisms, particularly in water bodies prone to cyanobacteria or eutrophication. On the other hand, reduced precipitation and reduced discharge due to industrial and agricultural development decreased the ability of rivers to dilute toxins. As a result, water-quality problems have led to more prominent water shortages in the southern, humid regions of China.

4.1.2 Cryosphere Hydrology

The impacts of cryosphere changes on the hydrology and ecology in the cold and arid regions of western China have significantly increased. Glacial runoff accounts for ~22 % of river discharge of inland river basins in western China, and the runoff is an important source of freshwater. Research has revealed that glacier runoff plays strong regulatory role in water availability when glacier coverage is over 5 % in a watershed. Glacial runoff produces more water in drought years and less water in humid years with lower air temperatures, therefore stabilizing arid oases.

Climate warming has led to increased glacier runoff in the last 50 years. Glacier runoff into the Tarim River and its main tributaries has significantly increased, with at least 1/3 of the increased river discharge coming from increased glacier runoff. The recent increased glacial runoff into the Yangtze River partly compensated for the decrease in the river's discharge. Intensification of glacier ablation and shrinkage has led to the areal expansion of some lakes that are mainly supplied by glacial runoff. In recent years, the frequency of glacier floods and glacial lake outburst floods has increased. Given differences of glacier coverage and glacier changes in each watershed, the impacts of glacial runoff on water resources still need further research, particularly the impact of the process of glacial runoff decrease after increasing during glacier retreat on river runoff needs refinement.

Permafrost degradation has a direct impact on basin's water cycle processes. Disappearance of permafrost, or thickening of the active layer, has caused a decline in the groundwater level (the water level above frozen soil), which results in decreased wetlands, rivers and lakes drying up, and the continued deterioration of the ecological environment. Permafrost degradation also affects the formation of basin groundwater systems, resulting in increased winter river discharge in many of the permafrost-covered regions.

Warming climate has led to a change in the distribution of annual river discharge, with an earlier start of the spring snowmelt process. The date of maximum monthly discharge(snowmelt discharge) of the Crane River(one tributary of the Altai Irtysh River) now occurs approximately one month earlier, with an increased discharge of ~15 %, and summer discharge has decreased in the past 50 years. Climate-related changes in accumulated snowfall can also affect the amount of snowmelt discharge.

4.1.3 Sea-Level Rise

Climate change has had a profound impact on the circulation and thermohaline environment of continental shelf areas of eastern China and the South China Sea. The temperature of coastal waters has shown a consistent warming trend from 1900 to 2008, with the warming rate of 1.3 ± 0.30 °C/100 years of the eastern continental shelf and northwestern Pacific, which is 2–3 times of the global ocean average warming, and this warming is closely related to a strengthening of the meridional heat transport of Kuroshio Current [a north-flowing ocean current on the west side of the North Pacific Ocean] (Wu et al. 2012). Another prominent phenomenon in Chinese coastal waters is the salinity of the Bohai Sea. The average salinity of the entire Bohai Sea has increased nearly 2 Psu (salinity unit) from 1961 to 1996. The salinity near the old estuary of the Yellow River in Bohai Bay has increased nearly 10 Psu, mainly caused by the reduction of Yellow River discharge to the Bohai Sea, and increased evaporation (Wu et al. 2004). The evaporation of the East China Sea has significantly increased, by ~1.2 cm/year, which has led to continuing loss of freshwater and heat, while the freshwater flux of the South China Sea has weakened, leading to decreased endothermic conditions from 1979 to 2008.

The average sea-level rise along Chinese coastal areas is 2.6 mm/year in the past 30 years, 0.8 mm/year higher than the 1.8 mm/year global average. Sea-level rise leads to aggravated coastal erosion, which has increased significantly since the 1950s. The majority of sandy shores, muddy shores, and coral reefs changed from deposition or stable states to erosional states, and total erosion increased. The average coastal erosion rate is 15–20 m/a from 1976 to 1982 in Hainan Island. Sea-level rise has increased coastal water depths, enhanced wave action, and intensified storm surges. Rising sea level has led to estuarine saltwater wedges upstream, increasing the intensity of seawater intrusion into rivers and aggravating groundwater salinization. This phenomenon is particularly evident near the deltas of major Chinese rivers. Seawater intrusion is more pronounced in coastal cities close to the Yellow Sea and Bohai Sea. Laizhou Bay, in Shandong Province, is one of the regions with the most serious seawater intrusion in China. The rate of intrusion has increased from 46 m/year during 1976–1979 to 404.5 m/year during 1987–1988 in this region. The sea-level rise has significant effects on salinity traced in Pearl River Delta. The traced distance of 250 mg/L of salinity line significantly increased with the upstream runoff frequency increases. The boundary of salinity traced back to the upstream significantly moved under certain upstream runoff conditions and the rise in sea level (Kong et al. 2010).

4.2 Terrestrial Ecosystems

Observational evidence increasingly shows that climate change has strongly influenced terrestrial ecosystems. The impacts of recent climate change are faster and more extensive than in any past period.

4.2.1 Forests

The structure, composition, distribution, phenology, productivity, and carbon sink behavior of forests have changed in response to climate change. The extent and optimal distribution range of the Larix, and spruce, fir, red cedar, and other tree species in the Xiaoxing'anling and eastern mountain ranges in Heilongjiang Province have moved northward from 1961 to 2003, due to climate change. The boundary of Larixgmeliniiin, the Daxing'anling moved northward approximately 1.5° latitude. The north and south boundaries of spruce–fir optimum distribution area moved northward latitude 0.5° and 1.5°. The optimum distribution area of pine moved northward 0.5° latitude (Liu et al. 2007). Climate warming has had a strong impact on the *Betula ermanii* (birch) tundra transition zone in the Changbai Mountains of northeast China, with the distribution from an altitude of 1900–1950 m migrated to 2150 m. The migration of entire populations of *B. ermanii* has recently shifted northward. Alpine meadows and some species in the forest transitional zone, in the Wutai Mountains of Shanxi Province, have shifted to northward with rising regional air temperatures. Shrubs have invaded alpine meadows, and the tree line has risen ~8.5 m per 10 years in the arid valleys of Yunnan Province (Moseley 2006).

The average spring temperature in China has increased by 0.5 °C, and spring has begun 2 days earlier, since the 1980s. Spring will start 3.5 days earlier if the average spring temperature rises by 1 °C. The start of spring would be delayed 4 and 8.8 days if the average spring temperature decreased by 0.5 and 1.0 °C, respectively. The observed start of spring in Shenyang (during 1960–2005) and in Beijing (1950–2004) suggests that the start of spring has occurred earlier with increasing spring air temperature.

4.2.2 Grasslands

Grassland areal extent has been degraded by 2 million hectares per year, mainly caused by droughts in northern China. The change of grassland ecosystems is more pronounced in the Qinghai–Tibet Plateau where the grassland ecosystems are more sensitive to air temperature increase. For example, grassland ecosystems with extreme vulnerability in the source regions of the Yangtze and Yellow rivers have continuously degraded since the 1960s. The area of high-cover alpine meadows and alpine swamp meadows has shrunk by 17.7 and 25.6 % (Wang et al. 2009). Meanwhile, the phenology of dominant plants of alpine meadows has undergone significant changes under climate change. For example, the flowering period of *Elymus nutans* grass, which is the dominant gramineous pasture in subalpine meadows, has advanced 10–14 days from 1985 to 2005, while the maturity period has advanced 20–24 days.

Water is the main limiting factor for grass growth in most parts of China. The productivity of the grassland ecosystem is mainly affected by precipitation. Increased precipitation improves soil moisture supply and enhances the photosynthesis rate, resulting in improved grassland productivity (Li et al. 2008). For example, the net primary productivity (NPP) of grasslands during the growing season in Inner Mongolia increased between 1982 and 2003, while the grassland productivity in typical cross-zone of agriculture and graze in northern China evidently increased. Reduced precipitation leads to declined grassland productivity. Such as in the pastoral districts of southern Qinhai Province and Gansu Province, middle of Ningxia, the Naqu district of Tibet, and pan regions of Qinhai Lake in Qinghai, the weaken hydrothermal tie has led to a general decline in forage yield with generally increased temperatures and decreased precipitation.

4.2.3 *Inland Wetlands*

The notable impacts of climate change on wetland ecosystems are mainly reflected in wetland hydrology, biogeochemical processes, plant communities, and wetland ecological processes. In the complex hydrology environmental systems with ice, frozen soil, snow, rivers, and lakes in the Tibetan Plateau, the area change of lakes is the combined effects of climate change, water, and heat. Regional lakes in the Tibetan Plateau, Mongolia, and Xinjiang have been significantly impacted by climate change, and some wetlands and lakes have expanded under climate warming (Ding et al. 2006). The area of wetland and lakes in Sanjiang Plain in northeast China, middle north, and east part of Tibet Plateau have sharply decreased during the same period, which mainly caused by increased air temperature and decreased precipitation. For example, the area of wetland in the source reach of Yangtze River and Yellow River and Ruoergai wetland have decreased above 10 %, among which the area of alpine swamp wetland in the source reach of Yangtze River has decreased about 29 % (Wang et al. 2009).

The composition and structure of wetland vegetation have significantly changed with climate change. The lobular camphor *Carex* has expanded in range to the central area of the wetland, while deepwater communities, such as *Carex lasiocarpa*, have declined in Sanjiang Plain in northeast China. The biological diversity of the wetland has sharply declined since the 1960s. For example, species of algae have decreased by 15.5 %, while the total number of algae has increased 181.4-fold, and fish species decreased by 44.4 % between the 1960s and the early 1990s in Baiyangdian wetland. Meanwhile, the eco-environment of surrounding degrades wetland turned deterioration. Desertification in the area surrounding the wetland has expanded in north China. Such as, the desert area around Hulun wetland has expanded to more than 100 km^2, and the degradation of pastures has accounted for a loss of more than 30 % of total pasture area in 1997. The wetland soil carbon pool is about 1/10–1/8 of the terrestrial soil organic carbon repository in China. The total loss over the past 50 years has probably reached 1.5 PgC.

4.2.4 *Biological Diversity*

Climate change and human activities have had a definite impact on biodiversity in China in the twentieth century. Some species have recently become extinct due to climate change and human activities. For example, Pantheratigrislecoqi, Equusprzewalskii, and Saigatatarica have become extinct in desert areas in China. Green peacocks were historically present in Hunan, Hubei, Sichuan, Guangdong, Guangxi, and Yunnan. They are currently found only in western, central, and southern Yunnan Province. Przewalski's gazelle used to range across of Inner Mongolia, Qinghai, and Gansu. They are currently distributed only in the Qinghai Lake region (Ma et al. 2006).

Climate change has affected the bird phenology and distribution of animals in the Qinghai region. Spot-billed ducks in the Bohai Sea area had been summer migratory birds before the 1990s, but have now become resident birds due to winter warming. Compared with the last century, 26 species of birds, including the bean goose, gray-headed blackbird, and bald Harrier, have disappeared from the Qinghai lake.

4.3 Terrestrial Environment

4.3.1 Land Use and Land Cover Change

The characteristic of land use and land cover change since the 1990s includes continued decreasing in arable land and continued increasing in forest area and the dynamic stability of grassland. The land use and land cover change greatly varied in different regions. Generally speaking, land use and land cover changes are the combination result of climate change and human activities. The rising temperatures, especially the relatively large rising temperature in north China, play an important role in the northward shift of margin for Chinese paddy and dry land boundaries. 87 % of new paddy fields concentrated in the northeast region. 59 % of newly reclaimed upland farmland located in three northeastern provinces and Inner Mongolia. If the future climate in the north China remains arid, the farmland area will be further reduced, while the area of woodland and pasture will be further increased (Liu et al. 2009).

The vegetation cover index NDVI in most regions of China had experienced an increase from 1982 to 2006, indicating that vegetation activity enhanced. The pronounced increase of NDVI with greater than 1 %/10a occurred including Beijing, Tianjin and its surrounding areas, southeastern Qinghai, and middle and west part of northern Xinjiang. NDVI in most areas, south of the Yangtze River decreased with more than −1 %/10ya. NDVI in eastern coastal areas slightly declined or did not change, while western regions experienced an increasing NDVI. The overall increase of NDVI in China is closely related to climate change. The main reason of increasing of NDVI is due to extending growing season and accelerated growth. Climate change, in particular the rising temperatures and summer precipitation, may be the main driving factor for an increase in NDVI. Temperatures are the master impact factor of increase in annual maximum NDVI relative to precipitation in the northeast China. The orders of impact of temperature on annual maximum NDVI for different types of vegetation decrease from forest, grassland, wetlands, and shrub to arable land. The projected global warming will have a significant impact on vegetation in northern China (Mao et al. 2012).

4.3.2 Desertification

Desertification processes are significantly affected by climate change. The relationships between typical climatic factors and land desertification are shown in Table 4.1. Increasing precipitation and the average temperature are extremely beneficial for the vegetation growth, and decreasing wind speed is also conducive to the transformation from semi-fixed sand dune to fixed dune. This is the main cause of some of the major sand-dune areas in China, such as in Horqin, and has reversed since the 1980s. Although air temperature has generally risen in the pastoral zones and sandy areas in northern China, increasing precipitation during the 10 years preceding the 1990s and a decrease of potential evapotranspiration, wind speed, and other elements have reversed the large-scale desertification in China by the start of this century.

Temperature is the most important factor affecting desertification in the Qinghai–Tibet Plateau. Desertification in the area has expanded in the past 50 years due to continued warming and permafrost degradation. The source regions of the Yangtze and Yellow Rivers in the Tibet Autonomous Region have the highest desertification rates in China. Desertification in the Qaidam Basin has also expanded (from 1961 to 2006), caused by warming climate and increasing high winds.

Climate change has exacerbated on the development of regional rocky desertification (a process of land degradation characterized by soil erosion and bedrock exposure. The impact of temperature changes is mainly through indirect effects of changes in vegetation, while precipitation intensity and precipitation amount have important direct impact on the rocky desertification. The main impacts of climate change on rocky desertification include the following: (1) Rocky desertification is increased in areas with more than 1200 mm annual precipitation in karst areas; the

Table 4.1 Correlation coefficients between climate elements and land desertification in typical sand-dune zones (after Li et al. 2009)

	Hunshandake sandy land			Horqin sandy land		
	Air temperature	Precipitation	Wind speed	Air temperature	Precipitation	Wind speed
Shifting sand dunes	0.761	−0.751	−0.833	0.684	−0.312	0.71
semi-fixed dunes	−0.15	0.829	0.324	−0.653	0.679	0.227
fixed dunes	−0.519	0.139	0.804	−0.624	0.658	−0.687

The results of Hunshandake sandy land come from the land use change from 1970 to 2000. The samples of Horqin sandy land are 31 greater the annual precipitation, the more serious the rock desertification. Annual precipitation has increased significantly in most eastern and northwestern parts of the karst areas in southwestern China, while precipitation in the middle and southwestern karst areas has significantly decreased. However, even in those regions with decreased annual precipitation, rainfall in flood season has increased, due to extreme rainstorms. Spatial patterns of precipitation are the main factors affecting the regional development of rocky desertification. Heavy rains in the karst mountain areas in southwestern China are more concentrated in the spring (~40 %) and summer (50 %) seasons for the cultivation of crops in these areas. (2) Climate change has caused vegetation degradation in the karst areas. The regional NDVI for most parts of northern Guangxi and eastern and southwest Yunnan provinces has decreased, accompanied by temperature increases over the past 20 years. The decrease in annual average NDVI is more pronounced in the northern Guangxi region. There is a significant negative correlation between temperature and vegetation index. The yearly average NDVI and NPP for evergreen and deciduous broad-leaved mixed forests, which are important in karst areas, have declined markedly in the recent past, causing rocky desertification.

4.3.3 Soil Erosion

Soil erosion is related to climatic conditions, the surface environment, and soil characteristics. Soil erosion in the Loess Plateau has a nonlinear relationship with annual precipitation, in that the erosion first increased and reached a peak with the increase in annual precipitation and thereafter decreased. Analysis of sediment data from 115 hydrological stations, and precipitation data from 276 rainfall stations in the Loess Plateau, suggested that the spatial and temporal variation of sediment erosion is consistent with that of precipitation changes over the past 50 years, and sediment erosion decreases with decreasing precipitation (Xin et al. 2009).

Precipitations in China have displayed distinct regional characteristics over the past 50 years. Significant climate warming and drying in the Loess Plateau has partly reduced the sediment load in rivers. However, predominant analyses indicate that water conservation contributed an average of 72.6 % to the decrease in soil erosion, while decreased precipitation contributed only 27.4 % in the Loess Plateau region (Xin et al. 2009). This suggests that precipitation changes caused by climate change are indeed affecting regional soil erosion. However, this effect is relatively small.

4.4 Agriculture and Forestry

Climate change has significantly affected agriculture and forestry in China on crop growth, development, and yield formation, as well as the spatial distribution of agricultural climate resources. The general increasing of accumulated positive temperature and the prolonged crop-growing season are benefit to expand thermophilic crops to high-latitude or alpine regions, changes of spatial distribution of different cropping system. As a result, the potential production has changed, agricultural biological and non-biological disaster has occurred more frequent, the agricultural production stability has decreased, lawn area has narrowed and moved northward, and the livestock productivity and the quality of livestock have declined. The fishery ecological environment has degraded, the traditional fishing grounds have disappeared, and fish reproductive capacity has decreased. The overall impact of climate change on structure, composition, function, productivity of high-latitude boreal forest is negative (Sun et al. 2010).

4.4.1 Agriculture

The changes of light, heat, water resources, and other factors in agriculture climate resource have great impacts on crop yield and quality. As a result of climate change, thermophilic crop zones in China have moved northward and into higher altitudes, the growing period has been extended, and the early-maturing varieties of crops have shifted to normal-maturing or late-maturing, which improved the adjustment of crop structure and layout. The crop belt has moved northward 150–200 km and 100–200 m vertically upward in mid-latitude zone due to climate warming. The national average days of accumulated temperature greater than 10 °C will extend 15 days when the annual average temperature increases by 1 °C. The average northern boundary of the cultivation of double crops and three crops a year during 1981–2007 has moved northward compared with that during 1950–1980 with increasing temperature and accumulated positive temperature (Fig. 4.2). The largest boundary of double crops in one year moving northward is in Shaanxi, Shanxi, Hebei, Beijing, and Liaoning provinces. The largest boundary of three crops in one year moving northward is in Hunan, Hubei, Anhui, Jiangsu, and Zhejiang provinces (Yang et al. 2010).

Climate change significantly impacts on crop potential production. Reducing the number of sunshine hours and precipitation will decrease the potential production, while temperature will increase the potential production. If the light and temperature conditions are suitable, the coordination between soil moisture and light and temperature conditions will help to achieve higher crop yields. On the contrary, the crop will slowly grow, and crop potential production will decline if the crop cannot get sufficient water supply. Decreasing precipitation at this century is the major factor of Chinese crop potential production. Climate warming is caused by the increase of atmospheric CO_2 and other greenhouse gas concentrations. As a substrate for photosynthesis, atmospheric CO_2 enrichment will lead to enhanced crop photosynthesis, which will help improve crop yields. However, the carbon content in the plant increases, the nitrogen content is relatively lower, the protein may reduce, and crop quality may decline. The CO_2 fertilization effect is a concrete manifestation of the crop growth environment and species breed to manage conditions.

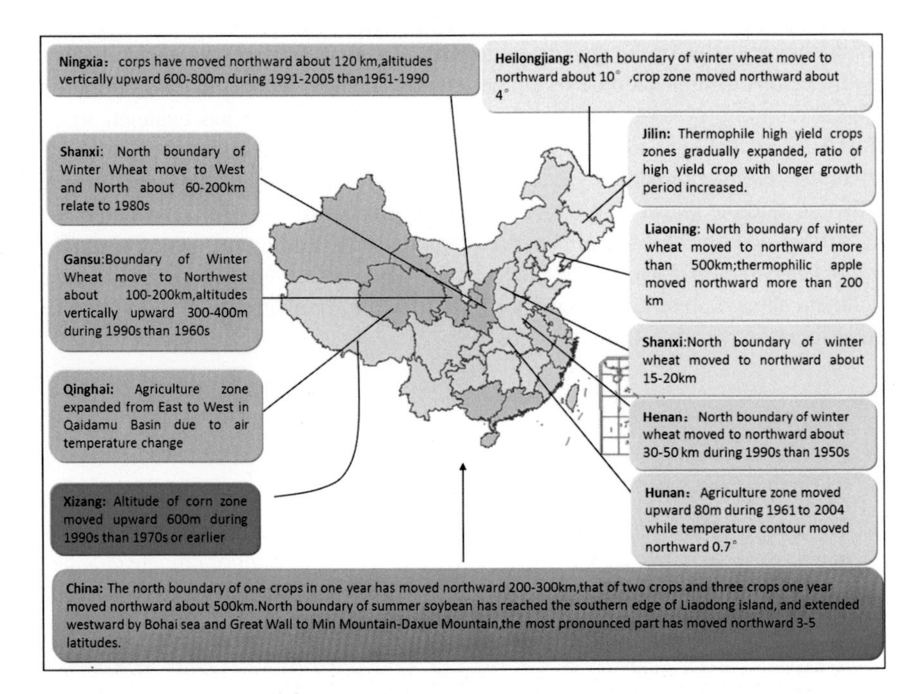

Fig. 4.2 The changes of planting boundary of crops in past 60 years in China (after Qin et al. 2012)

Currently, research is still not completely understood by the mechanism and extent of the stimulation of CO_2 on crop, and the research work in this area needs to be further strengthened.

Agricultural production in China will face three main challenges as a result of future climate warming. First, there will be an increase in agricultural production instability. Second, the geographic distribution and the structure of agricultural production will change. And third, agricultural costs will substantial increase (Ding 2003). Therefore, there is an urgent need to develop adaptation technologies and enhance China's adaptive capacity to mitigate the adverse effects of climate change and promote sustainable agricultural development.

4.4.2 Animal Husbandry

Pastoral areas in China are mostly located in arid and semiarid areas, or in cold regions in high latitudes; therefore, animal husbandry is sensitive to increases in population, land use changes, and in particular, to climate change. The impacts of climate change on animal husbandry are complex. Rising air temperatures and sustained or decreasing precipitation result in an increase in evapotranspiration of grasslands. High winds, dust storms, drought, and other extreme climate even negatively affect the growth of grass and result in a decline of pasture. More droughts strengthen the potential desertification of semiarid areas increased droughts in steppe regions, and longer droughts, which further reduce soil fertility, reduce the yield of grasslands, decrease the pasture cover, decrease the yield of forage, and decrease the grassland carrying capacity and stocking rates. On the other hand, climate warming, especially an increase in the average cold-season air temperature, favors the overwintering and reproduction of rats and other pests, leading to plant reduction and an increase in livestock epidemics. The impacts of climate change on livestock breeding both good and bad, but mainly, are adverse. Rising winter and spring temperatures and a reduction of snow-covered areas have reduced the snowstorm hazard in pastoral areas and benefited to livestock survival in winter and spring. However, higher air temperatures also lead to degradation of poultry production and an increase in mutated viral infections, which decrease livestock production.

4.4.3 Fisheries

Climate change has also impacted fisheries in China. Climate change undermines the stable structure of marine ecosystems, resulting in the destruction of ecosystems such as coral reefs, mangroves, and fish spawning grounds, which leads to a reduction of fisheries resources. Climate warming increases sea temperature, which directly affects the growth, feeding, spawning, migration, and mortality of fish. A reduction in, and disappearance of, sea ice can lead to the disappearance of spawning grounds for cold-water fish species, which inevitably affect their normal production and feeding behavior, making the endangered condition of these species worse. Climate change will lead to the fisheries population size and structure, and routes and times of fish migratory change dramatically, resulting in fisheries vanish or fisheries functions disappear.

4.4.4 Forestry

Climate change also has the important impact on forests, which is the main body of China's forestry. Climate change can affect the structure, composition, function, and productivity of forest ecosystems (especially high-latitude cold-temperature forests) and can threaten the recovery of degraded forest ecosystems. Rising temperatures in China have advanced the start of spring growth in woody plants, but there are spatial differences across the country. The start of spring has advanced in northeastern China, northern China, and the lower reaches of the Yangtze River, but it has been postponed in eastern southwest China, in the middle reaches of the Yangtze River, and in other areas. The extent of these changes across latitudes has decreased. Measurements suggest that the spatial distribution of some types of forests has changed in response to climate change, and the tree-line elevation has risen in some areas. Climate change has negatively impacted the secondary succession in some forests, and the recovery of tropical forest ecosystems, with an increase in the mortality of trees in secondary forest successions. The intensity and frequency of droughts have caused increased forest fuel accumulation, with prolonged periods of fire danger (now beginning in the early spring),resulting more early spring and summer fire-prone forest fires, and expanded geographic distribution area of forest fires, and then intensified frequency and intensity of forest fires. Climate change is also responsible for an increase in forest pests and diseases, an expansion of their spatial distribution northward, the advance of occurrence of forest pests, the increases of the number of generations, the shorten of cycles, the increase of scope and degree of harm, and promotion of the expansion and hazards of invasive alien pests.

4.5 Human Habitats and Health/Infectious Disease

4.5.1 Habitats

Climate change has direct effect on habitats. About 400 million people worldwide live within 20 km of a coastline or below an altitude of 20 km. Moreover, populations in developing countries are shifting to coastal cities due to the economic advantages of the coastal areas. Global warming raises sea level, increasing the risk that heavily populated coastal lowlands will be submerged by rising water levels. China is a maritime country, the mainland coastline is about 18,000 km, and there are more than 6000 islands. The coastal areas have been China's economically developed regions. The Pearl River Delta and Yangtze River Delta regions are relatively vulnerable regions.

The effect of urban heat island effect on habitats is more and more pronounced. The urban heat island effect is a phenomenon of increased temperatures and abnormal distribution of temperatures in urban environments. Many cities in China experience the urban heat island phenomenon. For example, the temperature difference between urban and rural areas in and around Beijing, Shanghai, and Lanzhou has increased in recent decades. Moreover, the urban heat island area has expanded with increasing urban temperatures.

Studies of annual average temperatures over the past 50 years suggest that urban heat in lands raise the annual average temperature, decrease the difference between inter-annual temperatures, and contribute to climate change. The average intensity of a heat island effect in China is less than 0.06 °C, which is close to the global average of 0.05 °C. Some studies have suggested that the average intensity of the heat island effect in China has risen by 0.1 °C per 10 years from the 1970s to the 1990s, while the average intensity in the Pearl River Delta has increased from 0.1 °C in 1983 to

0.5 °C in 1993. The areal extent of heat islands in major cities has generally increased. For example, the urban heat island in Shanghai has increased from 100 km^2 in the 1980s to 800 km^2 in the 1990s. Data collected from 1961 to 2005 indicate that the temperature rose more in urban belts than in non-urban areas, by 0.28–0.44 °C per 10 years, due to the rapid urbanization in the Yangtze River Delta region between 1992 and 2003. These differences are largest in summer and slightly smaller in autumn and spring. They are smallest in winter. The heat island effect had led to a 0.072 °C increase in the average temperature of the region between 1961 and 2005 and an increase of 0.047 °C from 1991 to 2005. The annual minimum temperature increased by 0.083 °C during the period from 1991 to 2005.

Climate change and the heat island have direct or indirect impact on people's living environment through different pathways. The comfort habitat of urban residents affects their health, work, and leisure life. Climate change and the heat island effect lead to urban climate anomalies and can lead to an increase in extreme weather events, such as the frequency and intensity of lightning and heavy rain. Urban heat island effects can also cause increased fog, droughts, and wind disturbances and ultimately lead to urban meteorological disasters.

The impacts of climate change on rural living are concerned. Agriculture as main field of climate vulnerable ecosystems, any degree of climate change will have potential or significant impact on agricultural production and related processes, thus affecting the lives of rural residents. Due to the dependence on the rural agriculture, forestry, and other climate-sensitive industries, therefore, rural residents are vulnerable to climate change impacts. Moreover, many rural areas are economically constrained. The ability to adapt to climate change in rural areas is limited. Poor areas and poor people are very vulnerable to climate change and extreme weather events.

Due to climate warming, soil erosion has intensified in China, leading to the desertification in rural areas which is becoming a serious problem. Decreasing precipitation due to climate warming leads to reduction in rain-fed agriculture, thereby changing the regional farming mode.

Climate change will increase the extreme weather events, leading to the increase of the frequency and intensity of floods and droughts. Due to poor infrastructure, the rural areas have lack of capacity to deal with extreme events. Climate change exacerbates environmental problems in rural areas. The already fragile rural infrastructure cannot guarantee normal life of rural residents when facing heavy rain, flash floods, snowstorms, and other extreme weather events.

4.5.2 *Life Facilities and Social Service*

An increase in the frequency and intensity of extreme weather events, such as lightning, heavy rain, and droughts, caused by climate change, leading localized flooding and road damage, traffic congestion, power outages, etc., seriously affects the normal operation of urban socioeconomic and urban infrastructure security. In recent years, meteorological disasters such as heavy rain, lightning, fog, drought, and other threats have become the focus of urban security objects.

Climate change has a greater social impact on some services sector. Such as, the insurance industry is closely related to climate and environmental change. Increases in the frequency and intensity of extreme weather events lead by climate change have increased the amount of insurance payments, thereby increasing the risk. China is one of the countries which is most severely affected by natural disasters in the world. China's insurance market is actually facing a far more serious catastrophe risk than international insurance market due to China's sustained and rapid economic development, increasing urbanization, population, and wealth. From another perspective, climate change also has brought strong demand to the property insurance, health insurance, and other fields. Adaptation to climate change is not only a complex challenge, but also a lot of opportunities to the financial sector.

4.5.3 Urbanization

Climate change affects resources and ecological environments within and around the cities, most prominently, water, thereby affecting the population carrying capacity and the development prospects cities. The direct effect of sea-level rise is flooding seawater. The salinization of groundwater has become an increasingly serious problem in many coastal cities in northern China, after continuous overexploitation of aquifers and recent sea-level rise. Sea-level rise also leads to coastal tidal inundation. Salt tide inundation reached an unprecedented peak in the dry season of 2005/2006, posing a serious threat to the security of the water supply in cities in the Pearl River Delta. Climate changes also lead to increased hurricane intensity and bring a serious threat to coastal urban life, property and urban economy, and transportation. Because the climate of northern and northeastern China has shifted to drier and warmer conditions since the 1980s, water demand in these regions has rapidly increased, leading to a growing gap between supply and demand of water resources, and thereby seriously restricted the development of many cities in northern China.

Increasing the frequency and intensity of extreme weather events, which caused by climate change, will lead to a variety of meteorological disasters intensified, especially droughts and floods will get worse, which is profoundly affected by climate change. According to statistics, the worldwide occurrence of major meteorological disasters (such as urban flooding, heat waves, urban haze, and lightning) in the 1990s is five times more than in the 1950s. Since the daily operation of its various functions must rely on the security of lifeline systems, such as transportation, electricity, telecommunications, water, gas, sewage, and other system, urban meteorological disasters will rapidly expand and spread to wide range or even entire city once the damage on lifeline systems come up.

4.5.4 Climate Change and Human Health

Global warming increases the frequency and intensity of regional heat waves, leading to an increase in heat-related injuries and diseases. As temperatures increase, fatigue, irritability, anger, and the number of accidents also increase, even the crime rate increases. Human brain tissue and myocardium are the most sensitive to high temperature with low pressure, prone to dizziness, impatient, irritability, etc., so as to cause some psychological problems (Table 4.2).

Climate change can lead to an increase in diseases in affected areas. The pathogens of many infectious diseases are sensitive to climatic conditions.

Increasing frequency and intensity of extreme events such as storms, floods, droughts, and typhoons will impact on human health through a variety of ways. These natural disasters cannot only directly cause casualties, or have indirect effects on health through damaged homes, migration, water pollution, food production (lead to hunger and malnutrition) etc., increasing the incidence of infectious diseases, but also damage the health service facilities.

Many pathogens, intermediary, the host and pathogen replication rate of infectious diseases are sensitive to climatic conditions.

The effect of climate change on waterborne diseases is complex, mainly due to socioeconomic factors that determine the supply of safe water. Extreme weather events, such as floods and droughts, probably increase the disease risk through contaminated water, poor sanitation, and other mechanisms.

The temperature and precipitation changes caused by global warming will affect the distribution pattern of diseases such as malaria and schistosomiasis. Climate change has both a direct and indirect impact on schistosomiasis transmission. When

Table 4.2 Climate change on human health effects (after Cao et al. 2001)

Effect on health	Climate change
Diseases caused by heat wave	Deaths related to cardiopulmonary disease increased with high or low temperature; death and heat-related illness increase during heat waves
Health effects of extreme weather and climate	Direct effects (deaths and injuries) and indirect effects (infectious disease, long-term psychological disease) of floods, landslides, landslides and storms; drought-induced disease or risk of malnutrition increase
Air pollution-related mortality and morbidity	Weather affect the atmospheric concentration of pollutants; weather affecting spatial distribution, seasonal changes, and generation of airborne allergens
Medium infectious disease	High temperatures increase development time of germs in carriers and increase the probability of the potential spread of the human body; spread of disease depends on special climatic conditions (such as temperature and humidity)
Skin diseases and eye diseases	Skin cancer and the incidence of various eye diseases increase with the increase of ultraviolet B (UV-B) radiation caused by the stratospheric ozone depletion

air temperatures rise, such as the extreme increase in the minimum air temperature in northern China where water imported to the region as part of the WTSN project, the snails that host the parasite may migrate northward. Similarly, if global average air temperatures increase 3–5 °C in the next century, the number of malaria patients could increase by 2 times in the tropics and more than 10 times in temperate zones, due to a spread of the host mosquito.

Dengue fever is an acute infectious disease caused by dengue virus which spread primarily by Aedes. It mainly explodes in tropical and subtropical countries and regions. Yi et al. (2003) found that the spread of dengue is mainly affected by mosquito density. The main meteorological factors which affect mosquito density are temperature and humidity, among which the temperature is the determining factors, indicating that the temperature is the determining factor of spread of dengue fever. Chen et al. (2002) suggested that the winter (3 months) temperature in the northern region of Hainan Province is not suitable for the spread of dengue fever, while the winter temperatures in the southern region of Hainan Province may be suitable for the spread of dengue fever. The dengue fever transmission in Hainan may have a fundamental change if the average temperature of winter months increases 1–2 °C. The periods of spread of dengue fever in the northern Hainan may extend to whole year, and the spread in the southern Hainan will be at a high level, thus making it possible to transform a non-endemic dengue fever in Hainan to a regional epidemic, resulting in the potential dangers of dengue fever that may be more serious.

Climate change may also cause the extinction of some species, and the generation of new species, including viruses and bacteria. Such as in the spring of 2003, the SARS virus diseases have been an outbreak in Guangdong, Beijing, Shanxi, and other places of China and bring great harm to society and people's health and life. The SARS outbreak is related to weather conditions. It is likely to always occur in the weather with inversion atmosphere, indicating that the climatic zones with inversion atmospheric are helpful to the SARS epidemic. The avian influenza occurred in the winter and spring, with a peak in January and February, while is rare in summer and autumn.

4.6 Other Economic and Social Areas

4.6.1 Industry

Climate change can affect the carrying capacity and environmental capacity of many natural resources that are used by industry, and extreme weather events can disrupt some industries. A reduction of water resources negatively impacts industries such as oil refinement, chemical production, fertilizer production, electric power production, metallurgy, mining, and textile production. The significant increase of lightning can adversely affect the electronic information industry.

Climate change also affects the patterns of energy and resource consumption. Higher air temperatures result in increased demands for water, cooling technologies, health care, recreation, etc., and decreased demands for winter goods and high energy consumption for water products. In order to reduce greenhouse gas emissions, the demand for clean energy further increases, which bring the development of relevant emerging industries.

Climate change further improved the requirements on transportation infrastructure and urban infrastructure. Increasing extreme weather further increases the likelihood in the destruction of these facilities, resulting the new growth of related industries. The blizzard in early 2008 in China was an extreme weather event that covered to 20 provinces (autonomous regions), resulting in the closure of more than a dozen airports, many highways, and the Beijing–Guangzhou Railway. Weather events such as these lead to logistics problems, higher consumer prices, and the emergence of other social instabilities.

4.6.2 Energy Production and Consumption

Energy consumption has significantly changed with climate warming. Energy consumption for cooling has increased, while energy consumption for heating in winter has decreased. Irrespective of climate change, population, per capita housing area, the increasing proportion of urban households has air conditioners inevitably increases growing residential cooling energy consumption. Studies on the energy consumption in major cities in Xinjiang Province in the past 40 years suggest that the number of days requiring heating has significantly declined, while the number of days requiring cooling has significantly increased. Climate models are predicting that the average global air temperature will rise by 1.4–5.8 °C by the end of the twenty-first century. The temperature in Xinjiang, especially the winter temperature, is projected to continue to rise. The cooling energy demand in hot season in most of the cities of Xinjiang is projected to continue increase, and heating energy demand in cold season is projected to continue to decrease.

The increase in frequency of extreme weather increases the emergency pressure on the energy system. Such as in 2008, snowstorm caused severe damage on power facilities in south China. The power outages were up to ten days in Chenzhou of Hunan Province. Meanwhile, snowstorm also led to serious losses on the communications industry. The direct economic losses of communications industry caused by the snowstorm were reported by 700 million yuan; about 14.2 million users were affected; many communications stations were affected; and 350 million yuan were invested to recover.

4.6.3 Tourism

The impacts of climate change on the tourism industry mainly are on regional tourism, tourism landscape, and tourism season. In China, tourism is mainly concentrated in the eastern coastal areas of the country. The number of days in which the temperature exceeded 40° increased significantly in recent years, and low cloud cover led to strong ultraviolet radiation, resulting in marine coral resources that have been significantly affected. On the other hand, rising temperatures have caused a rise in sea level over the past 50 years (with accelerated sea-level rise in recent years), resulting in an increased risk of coastal floods in seaside areas. Climate warming will change the composition, structure, and biomass of vegetation and wildlife species in China, leading to changes in distribution pattern of forest, biodiversity damage, and changes of some of the region's natural landscape and tourism resources, thus having impact on nature reserves, scenic spots, and forest parks which are based on biodiversity and natural ecosystems.

Climate change alters tourism and outdoor recreation business season, which is the interest of tourism enterprises. For example, in other areas of the country, decreased snowfall shortened the tourist season at some winter leisure resorts and caused some losses of tourism business which operating the snow and ice of winter recreation resort project. Warming air temperature severely affected marine coral resources, degrading local tourism resources and adversely affecting on tourism industry. Extreme events, such as storms, landslides, and mudslides, will directly effect on tourist traffic safety and health of visitors and even lead to accident personal injury or death, having adverse impact on the regional tourism industry.

Extreme weather events and increases in pests and infectious diseases all contribute to a decrease in the demand for travel and tourism. Changes on natural resources, ecological environment, and people's lifestyles cause by climate change will pose a serious threat on preservation of some human non-material cultural heritage, which are sensitive and vulnerable to climate change. Meanwhile, changes in biological phenology, landscape, and population activity patterns caused by climate change will profoundly affect the overall structure and layout of tourism areas and services, which may actually bring more tourism business opportunities. Summer resort tourism, ecotourism, and water sports may increase, and ice and snow tourism areas could be moved to higher latitudes and altitudes, which would preserve tourism business opportunities.

4.6.4 Major Projects

Climate change not only affects natural ecosystems and human living environment, but also has a profound impact on a large number of human constructions. Three Gorges Project, Water Transfer from South to North Project (WTSN), the Yangtze River estuary restoration project, Qinghai–Tibet railway, the Sino-Russian oil pipeline, the Three-North Shelterbelt Project and highways, high-speed railway, the south transmission lines, and other projects are major projects and have significant impact on the development of social and economic of China. It is very necessary to assess the impact of climate change on these projects, and in particular to evaluate the impact on the safe operation of these projects', possible threats and propose adaptation measures to climate change.

Three Gorges Project, the world's largest water conservancy project, is the key backbone project of governance and development of the Yangtze River, which has benefits to flood control, power generation, navigation, etc. Affected by climate change, the rainfall–runoff relationship in Yangtze River has changed. The probability of occurrence of drought in dry season and floods in humid season has increased, and the hydrological regime changed. Meanwhile, the annual runoff and hydrological process of the upper reaches of the Yangtze River have significantly changed under the rapid growth of water consumption with the economic development, water conservancy and hydropower project construction, inter-basin water transfer, and other factors. Under future climate change scenarios, the Three Gorges reservoir area is projected to significant warming and wet, and the frequency of flood, droughts, and other extreme events of the upstream region is projected to increase. The probability of occurrence of mudslides, landslides, and other geological disasters in the Three Gorges reservoir areas may increase with the increase of heavy precipitation, which will have adverse effects on the Three Gorges Project management, dam safety, flood control, flood fighting, etc. The increasing drought in the dry season will affect the water deposition, power generation, shipping, and water environment of Three Gorges Project. Thus, the early preventive measures are very important to adapt to climate change impacts. These measures include the following: by strengthening the joint scheduling with tributaries and reservoirs in upper and middle reach, flood diversion areas in middle and downstream reach and then reducing the risk of flood and insufficient power in the dry season; implementation of watershed eco-environment management and developing a long-term plan, strengthening infrastructure construction of flood control, drought, water supply, etc.; strengthening the monitoring efforts of geological and seismic disaster; increase in reservoir sewage treatments.

WTSN is a major strategic initiative to solve the severe water shortage in northern China and also an oversize large infrastructure project which is closely related to socioeconomic sustainable development of China. Climate change has little effect on the water volume of east and middle route of WTSN. However, it probably has adversely affected on water quality of east route of WTSN. Climate change, in particular, changes in precipitation, which probably has adverse effects on water resource and ecological environment of reservoirs of the middle route of WTSN. The middle route of WTSN probably encounters frequency of flood and drought conflict between the river networks in south and north China. Climate change may lead to the risk of water shortage in the source of WTSN, no water consume demand, water pollution, floods, earthquakes and other geological disasters. At the same time, climate change will not ease the water shortage situation in north China, while the receiving water in north China will add more soil moisture, which may lead to higher significant temperature, more latent heat and evaporation changes that may cause a certain degree of local climate change. Therefore, we need to take engineering measures and non-engineering measures to circumvent the possible project risks, hydrological risk, ecology and environmental risks, economic risks, social risks, etc. The basic starting point should be help to improve and restore the surface, ground, and coastal water environment and its associated ecosystems, while solving the short-term urgent needs of water shortages in yellow River, Huai River, and Hai River Basin. The long-term foothold of regional water issues in northern basin should focus on the basin themselves. We should ensure the necessary flow of the Yangtze River into the sea. We should coordinate inter-regional water allocation from the country's overall social, economic, and ecological interests.

The Yangtze River estuary restoration project is one of the most complex estuary management projects in the world. It is also the most ambitious water project in China. The channels, runoff into the sea, the coastal environment, and other issues caused by climate change should be assessing on focus. Climate warming led to sea-level rising, the sediment retention position of the Yangtze River estuary was traced upward, and waterway blockage was intensifying, while the impacts of storm surges, floods, heavy rain, and other extreme weather events on the Yangtze Delta restoration project become increasingly prominent. Therefore, we need to integrate measures including improving the seawall protection standards, comprehensive management on Yangtze Estuary Road, increasing coastal freshwater flow, and strengthen the Yangtze River environmental protection countermeasures.

Climate change has impacts on the Qinghai–Tibet Railway. The active layer thickness of permafrost has gradually deepened with the increasing air temperature along the Qinghai–Tibet Railway. The permafrost warming, melting, and degrading will change the stability of the basement of Qinghai–Tibet Railway. We must pay attention to the trend of future climate change along the Qinghai–Tibet Railway and the use of engineering and technical measures to adapt to the impacts of climate change. We need to build integrated management system with environment, health, safety, and transportation and to meet the needs of adaptive management.

Ecosystem degradation leads to soil and vegetation degradation, biodiversity reduction, exacerbating sand hazards, and other environmental disasters, having direct threat to West to East Gas Pipeline Project. Not only the increasing frequency of floods due to precipitation changes, but also the mountain landslides are major factors affecting the safety of the pipeline. The secondary disasters, such as ground subsidence, soil erosion, river sediment deposition, and so on, both would threaten the safe operation of gas pipeline. Strengthen the studies on the impact of climate change on project design parameters, such as ground temperature, and prevention of geological disasters are very important for engineering design, maintenance, and risk assessment.

China–Russia oil pipeline project is impacted by both the temperature changes and permafrost changes. Particularly, the underground ice melting led by climate change and anthropogenic factors, not only caused ground subsidence, but also led to the formation of secondary permafrost disasters, affects the stability of pipeline project.

We need to consider the freeze–melt disaster at the early stage, have an optimal design, and take appropriate measures to guarantee the pipeline deformation within the allowable range. Meanwhile, we need to strengthen the scientific monitoring, ensuring the stability and security operations of the gas pipeline.

Climate change has affected the vegetation distribution in "Three-North" Shelterbelt area, and the forest ecosystem structure will be changed. The single large tree species in "Three-North" Shelterbelt is expected to have more vulnerability under future climate change scenarios. Thus, there is urgent need for redrafting planning, optimal designing the structure of tree species, improving the quality of afforestation, combing with the development of the pattern of trees and shrubs with local climatic conditions, and then enhancing the ability to adapt to climate change.

Highways and high-speed rails are national transportation artery, facing with high rick of ice, snow, rain, fog, lightning, high temperature, and other inclement weather. Therefore, highways need to strengthen monitoring of meteorological information, timely adjustment of the construction and the operation plan. High-speed rails need to take measures to adapt to the impacts of climate change and focus on prevention of meteorological disasters in the key seasons and areas. South transmission line had been experienced serious loss in the 2008 snow and ice storms. Accompanied by the projected increasing in the probability of extreme weather events, south transmission lines need to strengthen prevention and protection on freezing rain, improve disaster warning, and other measures, enhancing the ability to adapt to climate change.

References

Cao, Y., Chang, X., & Gao, Z. (2001). Potential impacts of future climate changes on human health (in Chinese with English abstract). *Journal of Environment and Health, 18*(5), 321–315.

Chen, W., Li, C., Lin, M., et al. (2002). Study on the suitable duration for dengue fever (DF) transmission in a whole year and potential impact on DF by global warming in Hainan Province (in Chinese with English abstract). *China Tropical Medicine, 2*, 31–34.

Ding, Y. (2003). Climate change—facing to disasters and proulems (in Chinese with English abstract). *Natural Disaster Reduction in China, 2*, 19–25.

Ding, Y., Liu, S., & Ye, B. (2006). Climatic implications on variations of lakes in the cold and arid regions of China during the recent 50 years (in Chinese with English abstract). *Journal of Glaciology and Geocryology, 28*(5), 623–632.

Ding, Y., Ye, B., & Han, T., et al. (2007). Regional difference of annual precipitation and discharge variation over west China during the last 50 years. *Science in China Series D, 50*(6), 936–945.

Kong, L., Chen, X., Du, J. (2010). Impact of sea-level rise on salt water intrusion based on mathematical model. *Journal of Natural Resources, 25*(7), 1097–1104.

Li, X., Han, F., Zhang, C., et al. (2009). Influence of climate change on mosaic landscape of sand land-wetland in middle-east Inner Mongolia (in Chinese with English abstract). *Chinese Journal of Applied Ecology, 20*(1), 105–112.

Li, G., Zhou, L., & Wang, D. (2008). Variation of net primary productivity of grassland and its response to climate in Inner Mongolia (in Chinese with English abstract). *Ecology and Environment, 17*(5), 1948–1955.

Liu, D., Na, J., Du, C., et al. (2007). Changes in eco-geographical distributions of major forestry species in Heilongjiang Province during 1961–2003 (in Chinese with English abstract). *Advances in Climate Change Research, 3*(2), 100–105.

Liu, J., Zhang, Z., Xu, X., et al. (2009). Spatial patterns and driving forces of land use change in China in the early 21st century (in Chinese with English abstract). *Acta Geographica Sinica, 64*(12), 1411–1420.

Ma, R., & Jiang, Z. (2006). Impacts of environmental degradation on wild vertebrates in the Qinghai Lake drainage, China (in Chinese with English abstract). *Acta Ecologica Sinica, 26*(9), 3066–3073.

Mao, D., Wang, Z., Luo, L., et al. (2012). Correlation analysis between NDVI and climate in northeast China based on AVHRR and GIMMS data sources (in Chinese with English abstract). *Remote Sensing Technology and Application, 27*(1), 77–85.

Moseley, R. K. (2006). Historical landscape change in northwestern Yun-nan, China. *Mountain Research and Development, 26*, 214–219.

Qin, D., Ding, Y., & Mu, M. (Eds.). (2012). *Evolving climate and environment in China: 2012 (II): Impacts and vulnerability (in Chinese)*. China: China Meteorological Press.

Su, H., Shen, Y., Han, P., et al. (2007). Precipitation and its impact on water resources and ecological environment in Xinjiang region (in Chinese with English abstract). *Journal of Glaciology and Geocryology, 29*(3), 343–350.

Sun, Z., & Wang, C. (2010). Impact of changing climate on agriculture in China (in Chinese with English abstract). *Science and Technology Review, 28*(4), 110–117.

Wang, G., Li, N., & Hu, H. (2009). Hydrologic effect of ecosystem responses to climatic change in the source regions of Yangtze River and Yellow River (in Chinese with English abstract). *Advances in Climate Change Research, 5*(4), 1673–1719.

Wu, D., X., Mu, L., Li, Q., et al. (2004). The long term of salinity in the Bohai Sea and its major control element (in Chinese). *Advances in Natural Science, 14*(2), 191–195.

Wu, L. X., Cai, W. J., Zhang, L. P., et al. (2012). Enhanced warming over the global subtropical western boundary currents. *Nature Climate Change, 2*, 161–166.

Xin, Z., Xu, J., & Yu, X. (2009). Temporal land spatial variability of sediment yield on the Loess Plateau in the past 50 years (in Chinese with English abstract). *Acta Ecologica Sinica, 19*(3), 1129–1139.

Yang, X., Liu, Z., & Chen, Z. (2010). The possible effects of global warming on cropping systems in China I: The possible effects of climate warming on northern limits of cropping systems and crop yields in China. *Scientia Agricultura Sinica, 43*(2), 329–336.

Yi, B., Zhang, Z., Xu, D., et al. (2003). Correlation between dengue fever epidemic and climate factors in Guangdong Province (in Chinese with English abstract). *Journal of Fourth Military Medical University, 24*(2), 143–146.

Zhang, J., Zhang, S., Wang, J., et al. (2007). Study on runoff trends of the six larger basins in China over the past 50 years (in Chinese with English abstract). *Advances in Water Science, 18*(2), 230–234.

Chapter 5
Projections of Climate Change and Its Impacts

Wenjie Dong, Mu Mu and Xuejie Gao

Abstract Results from multiple CMIP5 model ensembles projected a warmer climate in China with a general increase of precipitation in the future. Under the emission scenario of RCP8.5, a warming of 5.1 °C and 13.5 % increase of precipitation is projected in the end of the twenty-first century. However, spread areas with decreased precipitation are simulated by the higher resolution regional climate model. Climate change leads to the northward shift of the crop cultivation boundaries. The yield of

The following researchers have also contributed to this chapter

Shaohong Wu

Institute of Geographic Sciences and Natural Resources Research, Chinese Academy of Sciences, Beijing 100101, P.R. China

Baisheng Ye

Cold and Arid Regions Environmental and Engineering Research Institute, Chinese Academy of Sciences, Lanzhou 730000, P.R. China

Yongjian Ding

Cold and Arid Regions Environmental and Engineering Research Institute, Chinese Academy of Sciences, Lanzhou 730000, P.R. China

Yong Luo

Center for Earth System Science/Institute for Global Change Studies, Tsinghua University, Beijing 100084, P.R. China

Ying Shi

National Climate Center, China Metrological Administration, Beijing 100081, P.R. China

W. Dong (✉)
State Key Laboratory of Earth Surface Processes and Resource Ecology, Beijing Normal University, Beijing 100875, People's Republic of China
e-mail: dongwj@bnu.edu.cn

W. Dong
Zhuhai Joint Innovative Center of Climate-Environment-Ecosystem, Beijing Normal University, Zhuhai 519087, People's Republic of China

M. Mu
Institute of Oceanology, Chinese Academy of Sciences, Qingdao 266071, People's Republic of China
e-mail: mumu@qdio.ac.cn

D. Qin et al. (eds.), *Climate and Environmental Change in China: 1951–2012*,
Springer Environmental Science and Engineering,
DOI 10.1007/978-3-662-48482-1_5

wheat, maize, and paddy rice will decrease if the CO_2 fertilization is not considered. Northward shift of different types of forest and significant reduction of the tundra coverage over the Tibetan Plateau are expected. Although increases in the surface runoff depth are found in many areas, the shortage of water resources in the Haihe and Yellow River basins will be hard to be relieved. Sea-level rise due to the warming, plus the relative rise due to the over-extracting of ground water, will thread the local ecosystems as well as the social and economic society.

Keywords Climate models · Climate change projection · Climate change impact · Agriculture · Vegetation · Water resources

5.1 Progresses and Validation of Climate System Model

5.1.1 Progresses of Climate System Model

Under the global warming, how the future climate will change is one of the major concerns amount scientists, the public, and policy makers. More attention has been paid to the climate change projection in the timescales from decades to a century, since it is closely related to the planning of social and economic development.

Climate system models are the primary tools in simulating and projecting climate change. They are the mathematical expression of the climate system. Their simulations are conducted in supercomputers through computer programs. The reliability of the projections is firstly based that they are established on the common physical laws (such as quality, conservation of energy and momentum), as well as a large number of observational facts. Secondly, their reliability comes from the capabilities of reproducing major characteristics of the present-day climate, such as the large-scale distribution of atmospheric temperature, precipitation, radiation and wind, and ocean temperature, ocean stream, and sea ice cover, as well as the evolution of important climate system, e.g., the monsoon, seasonal change of temperature, storm tracks, rain belt, and the hemisphere-scale oscillation in extra-equatorial area, simulation capabilities on which indicate that the models can describe the basic physical processes of climate, thereby giving the confidence for climate change simulations. Finally, the capabilities on simulating and reproducing the major characteristics of paleo- and present-day climate change also proved their confidence in the climate change projection. Uncertainties in climate change projection arise from the emission scenarios of greenhouse gases in the future, the error

X. Gao (✉)
Institute of Atmospheric Physics , Chinese Academy of Sciences, 9804, Chaoyang District, Beijing 100029, People's Republic of China
e-mail: gaoxj@cma.gov.cn

on reproducing climate system because of the limitations in current capabilities of climate models, and the internal variability in models and climate system.

Compared with numerical weather prediction model, climate system models mainly focus on the interaction among the different spheres of the climate system, and the effects of anthropogenic activities. The major model components consist of atmosphere, land surface, ocean, sea ice, aerosol, carbon cycle, vegetation, and atmospheric chemistry. These model components are developed and modified separately, and then coupled together, forming the climate system model. The development of a model is closely relied on the gradually increasing understanding and recognition on physical, chemical, and ecological processes and their interactions. Continuously upgrade of computer capabilities provides with the basic requirements for the development of the model, and therefore, the model becomes more complex and more complicated. By now, besides the coupling of atmosphere–land surface–ocean and sea ice, the climate system models also couple with the aerosols and carbon cycle processes. Dynamic vegetation and atmospheric chemical processes have also stated to be coupled to the climate system model gradually. In the fifth assessment report of IPCC (IPCC AR5) (IPCC 2013), there is special attention on the effects of global carbon cycle on climate variability and climate change, try to better understand the land and sea absorption capacity on the greenhouse gas as well as how it will influence and respond to climate change in the future.

After 20 years of development and improvement, the multi-spheres-coupled model already has significant progresses. Their simulations have become the scientific basis for analysis and providing climate change projection which contributes to the IPCC climate change assessment reports. From the first to the fourth assessment reports of IPCC, a total of 11, 14, 34, and 26 climate models are used, respectively, with gradual improvements in physical process and resolution in the models.

Six climate system models developed in China have participated in Coupled Model Intercomparison Project Phase 5 (CMIP5) in supporting of the IPCC AR5. They are FGOALS-s2, FGOALS-g2, BCC_CSM1.1, BCC_CSM1.1_M, BNU-ESM2.0, and FIO-ESM, respectively.

FGOALS-s2 is developed by the State Key Laboratory of Numerical Modeling for Atmospheric Sciences and Geophysical Fluid Dynamics of the Institute of Atmospheric Physics, Chinese Academy of Sciences (LASG/IAP) (Zhou et al. 2008). Its atmospheric component is the 2nd version of Spectral Atmospheric Model of IAP/LASG (SAMIL2.0). The oceanic model is LICOM2.0, and land component model is the common land model (CLM3). The sea ice model is CSIM4 which includes the consideration of both the thermodynamics and dynamic processes. FGOLAS is coupled directly the ocean and atmosphere component without flux adjustment through the coupler of NCAR CPL6.

FGOALS-g2 is jointly developed by LASG/IAP and Earth System Research Center of Tsinghua University. Its coupler, ocean model, common land model are the same as FGOALS-s2, except the atmospheric component uses GAMIL2.0 and the sea ice model is CICE_LASG.

BCC_CSM1.1 (BCC_CSM1.1_M) is developed by National Climate Center, Chinese Meteorological Administration, and its atmospheric model is

BCC_AGCM2.0.1, land model is BCC_AVIM1.0.1, the ocean circulation model and sea ice model are MOM4-L40 and SISM, respectively. The model also includes global carbon cycle and dynamic vegetation processes (Wu et al. 2010).

BNU-ESM2.0 is developed by Beijing Normal University. It is based on its common land model CoLM, with the atmospheric model of CAM3.5, ocean model MOM4pl, and sea ice model CICE4.1 coupled by the coupler of CPL6. The model also includes the carbon cycle over land and ocean.

FIO-ESM is from the First Institute of Oceanography, State Oceanic Administration of China. It is developed from CCSM3.0, with the coupling of its own wave model, as well as inclusion of carbon cycle module in the atmosphere, ocean, and land models.

5.1.2 Validation of the Multi-model Simulation Over China

In this section, the performances of multi-model ensembles in simulating the present-day (1986–2005) climate over China are evaluated by comparing the twentieth-century climate in coupled model (20C3M) simulations against observation (Dong et al. 2012). The focus is on the annual mean temperature and precipitation. The models we used include BCC-CSM1-1 (China, National Climate Center), CanESM2 (Canada), CNRM-CM5 (France), FGOALS-s2 (China, Institute of Atmospheric Physics), GISS-E2-R (USA), MIROC5 (Japan), MIROC-ESM-CHEM (Japan), MIROC-ESM (Japan), and NorESM1-M (Norway).

Temperature is an important factor for the climate for a certain region, which can be used to describe the degree of warming or cooling, and to measure the heat balances. It is also an important climatic factor in influencing or representing the local ecological systems and the environment. As can be found in Fig. 5.1a, the temperature in eastern China is characterized by a colder north and warmer south. In southern coast, the temperature is warmer than 20 °C, and in the northeastern part of the northeast China, it is below −5 °C. In western China, the temperature is largely affected by altitude and the value of ranges from 10 to 15 °C in the basins and from −10 to −5 °C in northern part of the Tibetan Plateau. The model shows good performances in reproducing both the pattern and values of the present-day temperature. Spatial correlation coefficient between the model simulation and observation is of 0.96, and the average bias is −1.0 °C. The difference between the simulation and observation is in general within ±1 °C in eastern China except over the mountain areas. Larger bias is found in western China with the warm bias in the mountainous regions and cold bias in the basins exists. In addition, a general bias is found over the Tibetan Plateau. These cold or warm biases are mostly due to the smoothed topography by the low resolution of the models.

The observed precipitation shows a decrease from south to north and northwest, with maxima greater than 5 mm/d found in southeast China, and the minimum of less than 0.5 mm/d in northwestern China. In northwestern China, greater precipitation is found over the high mountains compared to the surrounding areas (Fig. 5.1c). The

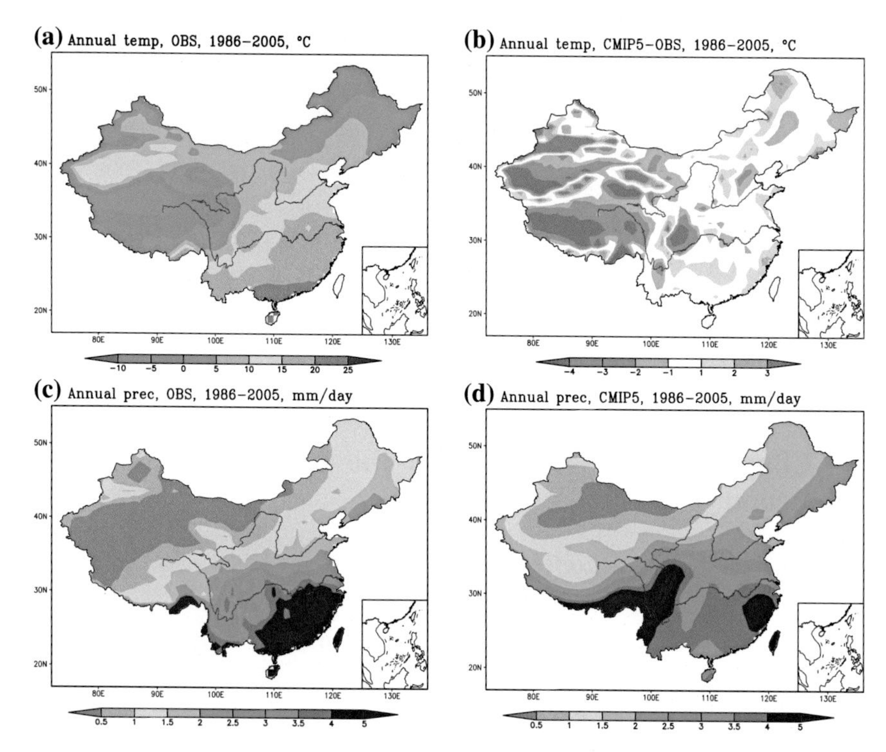

Fig. 5.1 Annual mean temperature (°C) and precipitation (mm/d) from 1986 to 2005 **a** observed temperature; **b** temperature differences between simulation and observation; **c** observed precipitation; **d** precipitation differences between simulation and observation

models can reproduce the general spatial distribution of precipitation with the spatial correlation coefficient between the simulation and observation is 0.81, while the simulated precipitation is 45 % more than that observed. However, as can be found in Fig. 5.1d, an artificial precipitation center in the eastern edge of the Tibetan Plateau exist, and the models failed in reproducing the topographic precipitation over the mountains in northwest China. These deficiencies are similar to the previous global models (e.g., CMIP3, see Xu et al. 2010), which is also due to the coarse resolution of the models (Gao et al. 2006, 2008). Further more, underestimation of precipitation in south China is found in the simulations.

Regional mean changes of temperature and precipitation over China in twentieth century are shown in Fig. 5.2a, b, respectively. As can be found in the figure, the model can reproduce the warming trend well. The linear trend of warming over China from 1961 to 2005 is 0.28 °C/10a from the observation, while the simulated is slightly lower as 0.20 °C/10a. The correlation coefficient between the simulation and observation is 0.82. The linear trend of precipitation in the simulation is 0.09 %/10a, while that in the observation is of 0.65 %/10a, and the correlation coefficient between simulation and observation is only −0.13. Compared to temperature, the model capacity in simulating precipitation trend is poor.

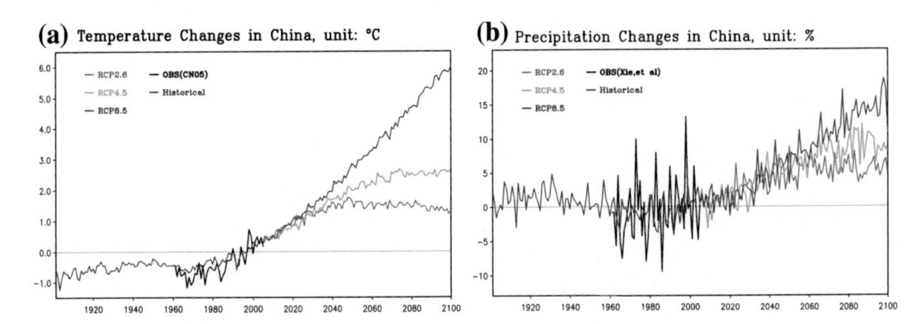

Fig. 5.2 Regional mean temperature (°C) and precipitation (%) change over China (in relative to 1986–2005) **a** temperature; **b** precipitation (*black line* is observation; *purple* is historical simulation; *red, green,* and *blue* are simulations under RCP2.6, RCP4.5, and RCP8.5)

5.2 Climate Change Over China in the Twenty-First Century

5.2.1 CMIP5 Multi-model Projections

In general, the ensemble mean simulation of global mean surface temperature changes by CMIP5 is consistent with that by CMIP3. And the consistencies among different models are also high. Focusing on China region, the temperature and precipitation temporal changes under RCP2.6, RCP4.5, and RCP8.5 emission scenarios are shown in Fig. 5.2. The values of regional mean change at three periods of the twenty-first century (2011–2040, 2041–2070, and 2071–2100) are presented in Table 5.1. We can found that the warming of the three scenarios in the earlier half of the twenty-first century is similar to each other, indicating the less dependency of scenarios in the earlier periods. The temperature will decrease under RCP2.6 emission path first, with the warming trend values of 0.27 °C/10a from 2011 to 2050 and then changed to −0.04 °C/10a from 2051 to 2100 following the pathway of the emission. The warming will continue until 2070 (0.34 °C/10a) under the RCP4.5 scenario, and subsequently begin to stabilize. Persistent warming can be found in the whole century in the RCP8.5 situation, with the regional mean temperature increase by 5.1 °C in the end of this century, from 2071 to 2100. The regional mean precipitations under the three scenarios are in gradual increase from 2011 to 2100. The increasing trends are 0.5 %/10a, 1.1 %/10a, and 1.8 %/10a under RCP2.6, RCP4.5, and RCP8.5 scenarios, respectively. The precipitation of the twenty-first century will increase by 13.5 % under the RCP8.5 (Table 5.1).

Let us take RCP8.5 as an example, Fig. 5.3 shows the spatial distributions of annual mean temperature and precipitation changes during three periods of the twenty-first century (2011–2040, 2041–2070, and 2071–2100). It can be observed from the figure that the changes of both become larger following the time evolvements. During 2011–2040, the temperature increases over the whole region, with less increase at the southern coast and southwest, with the values of 0.5–1 °C, and

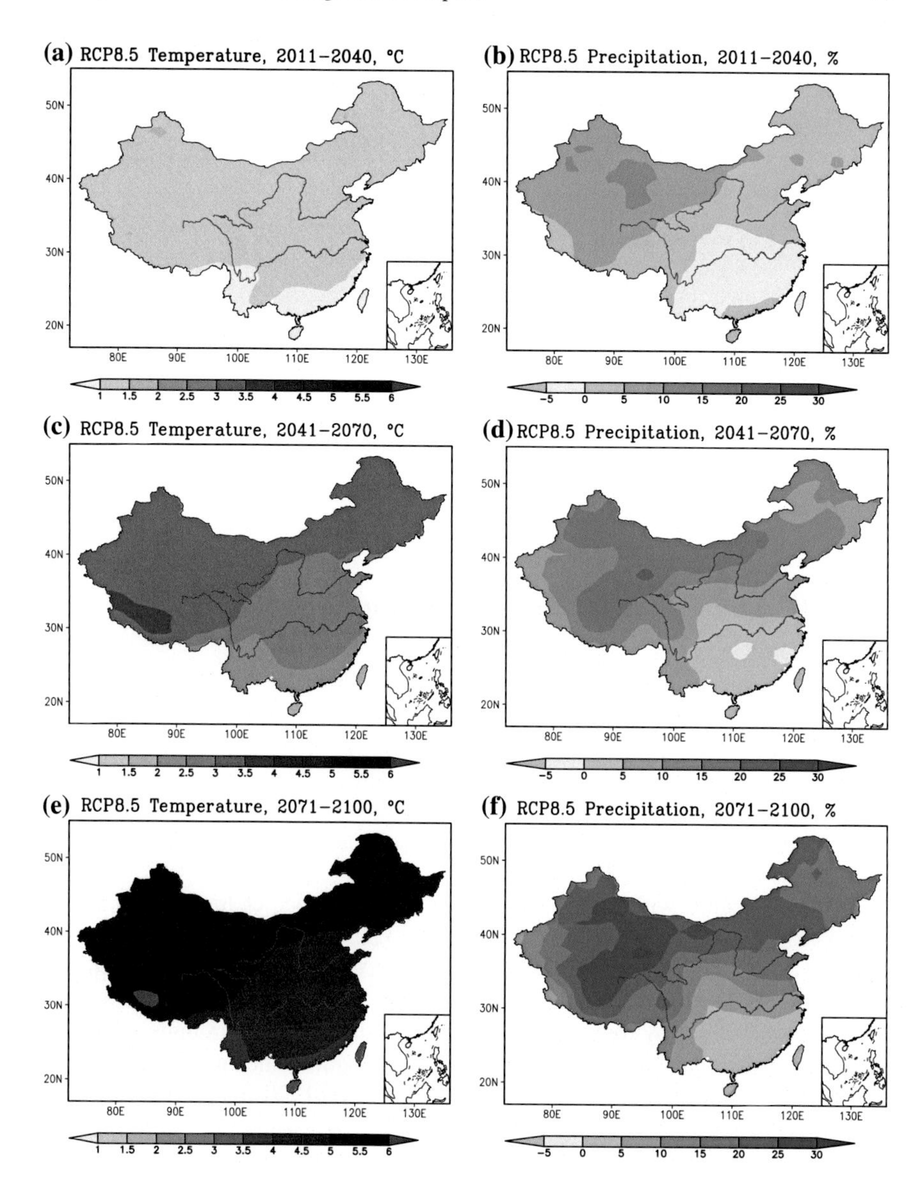

Fig. 5.3 The regional annual mean temperature (*left*, °C) and precipitation (*right*, %) changes over China from 2011 to 2040 (*top*), 2041 to 2070 (*middle*), and 2071 to 2100 (*bottom*), under RCP8.5 scenario (relative to 1986–2005)

1–1.5 °C found elsewhere. For precipitation, an increase can be up to 10 % is found in the northwest, while decreased precipitation is found south of the Yangtze River. In the period of 2041–2070, temperature will rise between 2 and 2.5 °C in southern China and 2.5–3 °C in eastern China and central China. The most remarkable increase is found in Tibetan Plateau, with the value ranging from 3.5 to 4 °C. Precipitation will

Table 5.1 Mean temperature and precipitation changes in different periods of the twenty-first century under different emission scenarios

	Temperature (°C)			Precipitation (%)		
	RCP2.6	RCP4.5	RCP8.5	RCP2.6	RCP4.5	RCP8.5
2011–2040	1.1	1.1	1.2	2.4	2.3	2.3
2041–2070	1.5	2.1	3.0	5.2	6.8	8.1
2071–2100	1.4	2.5	5.1	5.6	8.8	13.5

increase across the country, including a less than 5 % increase over south China, and 15–20 % increase in the northwest China and central Tibetan Plateau. The spatial distribution of the temperature and precipitation increase of 2071–2100 is similar to the period of 2041–2070. The largest warming can be found in the northeast, north of northwest and Tibetan Plateau, with the amplitude of more than 5.5 °C. Northwest China is also the place where greater precipitation increase of over 50 %.

5.2.2 *Regional Climate Model Projections*

Resolution of the global climate models is usually low due to their complexity, longer simulations of centuries, thus the large requirement of computer resources. In regional and local scale, climate change projection can be derived by dynamical or statistical downscaling. The dynamical downscaling uses the observed or coarse resolution global model results as the initial and boundary conditions to drive the higher resolution climate model (topically the regional climate models) to perform the simulations at the global or regional scale.

China is a region with complex topography and surface characteristics and is located in the East Asian monsoon area. The capability of the high-resolution regional climate models shows better performances compared to the driving global models. In recent years, several regional climate change simulations have been conducted over China by regional climate models, such as a 20-km-high-resolution simulation performed by the RegCM3 nested within the global model of FvGCM/CCM3 (Gao et al. 2008). Two periods, 1961–1990 for present-day, and 2071–2100 for the future climate under IPCC SRES A2 scenario, have been conducted in the simulations (hereafter referred as FvGCM-RegCM simulation).

Temperature is strongly dependent on elevation. Since the regional models can provide with more detailed description of elevation, it can usually simulate the spatial distribution of the surface temperature better. For precipitation, global models tend to simulate an precipitation center over eastern edge of the Tibetan Plateau in summer and annual mean, while the observed rainfall center is located at south of the Yangtze River. FvGCM simulation shows the same characteristics, while the RegCM3 simulation shows a large improvement. The major rainfall center in RegCM3 simulation located mostly at the south of the Yangtze River, and the artificial rainfall center in the eastern edge of the Tibetan Plateau has also been

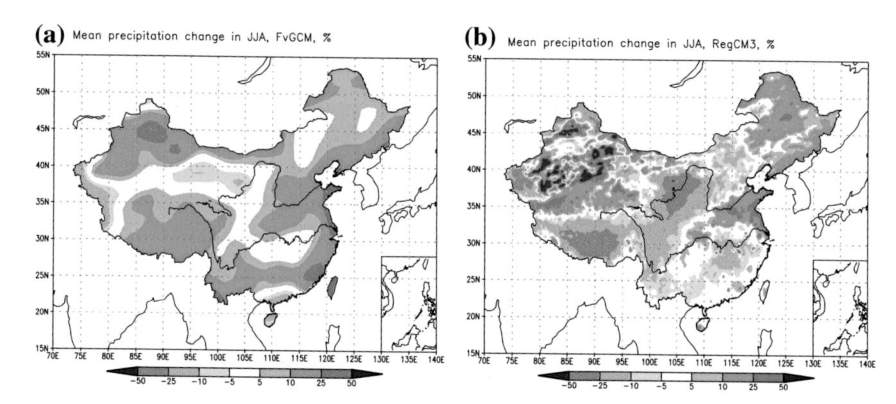

Fig. 5.4 Precipitation change in JJA in 2071–2100 over China (relative to 1961–1990, %) **a** projection by FvGCM; **b** projection by RegCM3

removed, thus agree with the observation better. In addition, RegCM3 reproduces the topographic precipitation very well, which is also consistent with the observations. For example, the relatively lower precipitation over the Qilian Mountain and higher precipitation over Qaidam basin can be well simulated by RegCM3 but not FvGCM (figures not shown for brevity).

Future climate change projections from regional climate models show differences with the driving global model, especially for the precipitation. The warming simulated by FvGCM and RegCM3 is of 3.7 and 3.5 °C, respectively, however large differences concerning the spatial and seasonal distributions. The annual mean precipitation increase simulated by FvGCM is 11.3 %, while that by RegCM3 is only of 5.5 %. The two models show similar spatial pattern in the winter simulation, although the regional model provides with larger spread of the decreased areas and greater values of the decrease. The two models showed large differences in summer (Fig. 5.4a, b). Precipitation projection by FvGCM is in general increase which agrees with most of the other global models (e.g., CMIP5), while the RegCM presents decrease in most areas, especially in the middle and upper reaches of the Yellow River, Tibetan Plateau, and the south China, with the largest value of decrease up to 25 %. Analysis has shown that this difference is mainly caused by the higher resolution in regional climate model, which provides more realistic and larger topographic forcings, resulting different circulation, and water vapor transportation changes.

5.3 Climate Change Impacts Over China in the Twenty-First Century

5.3.1 Impact on Regional Agricultural Production

Agriculture is one of the most sensitive sections. Climatic change will exert potential or obvious effects on agricultural production and its related processes (Xiong et al. 2006; Yang et al. 2010, 2011).

1. **Agro-climate resources**

Compared with the basic climate period (1961–1990), under SRES A2 and B2 climate scenarios, during the period in which daily average air temperatures steadily pass through 0, 3, 5, 10, and 15 °C, the lasting days, accumulative temperatures, average free frost period, solar radiation, precipitation, and potential evapotranspiration in most parts of China will obviously increase, however, the regional difference is obvious. The obvious change of agro-climate resources in China would affect potential agriculture productivity.

2. **Crop cultivation system**

Northern boundary of zero-class crop cultivation: Compared with the period of 1950s–1980s, under SRES A1B climate scenario, the northern boundaries of two cropping per year and three cropping per year in China will move northward, and some parts of one cropping per year in east Inner Mongolia and the border of Liaoning province on Inner Mongolia will change into the region of two cropping per year.

Northern boundary of wheat crop cultivation: Compared with the period of 1950–1980, under SRES A1B climate scenario, the northern boundaries of winter wheat cropping in northern China will move northward and westward. The northern boundary will move northward to the line from Heishan, through Anshan, Youyan to Dandong, during 2011–2040, and will continue to move northward with the distance of about 200 km in the east part and about 110 km in the west part during 2041–2050 (Fig.5.5).

Northern boundary of maize crop cultivation: Compared with the period of 1961–1990, under SRES A2 and B1 climate scenarios, the northern boundaries of early-, middle-, and late-maturity maize varieties in northeastern China will move northward. The early-maturity maize varieties within the maize cultivation region might be replaced by middle- and late-maturity maize varieties under climate change.

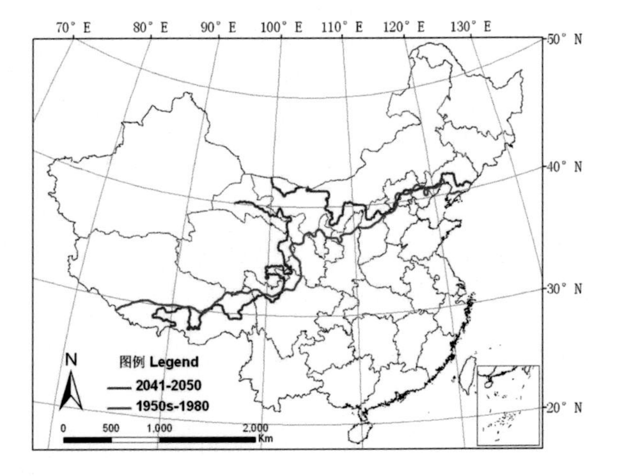

Fig. 5.5 Northern boundary of wheat crop cultivation in the present day and 2041–2050 under SRES A1B scenario over China

3. **Crop yield**

Wheat: Under SRES A2 and B2 climate scenarios, the yield would decrease averagely about 21.7 and 12.9 % for the rain-fed wheat, and 8.9 and 8.4 % for the irrigation wheat. However, the yields of both rain-fed and irrigation wheat would increase with the consideration of CO_2 fertilization.

Maize: Under SRES A2 and B2 climate scenarios, the yield of rain-fed maize would increase; however, the yield of irrigation maize would decrease with the consideration of CO_2 fertilization. Under A2 climate scenario, the yield would increase 20.3 % for the rain-fed maize and decrease 23.8 % for the irrigation maize. Under B2 climate scenario, the yield would increase 10.4 % for the rain-fed maize and decrease 2.2 % for the irrigation maize.

Paddy rice: Under SRES A2 and B2 climate scenarios, the yield in the middle and lower reaches of Yangtze River would decrease about 15.2 and 15.0 % for single paddy rice without the consideration of CO_2 fertilization and 5.1 and 5.8 % for paddy rice with the consideration of CO_2 fertilization, respectively (Yang et al. 2010).

5.3.2 Impact on Regional Vegetation

Climate change (climate scenarios SRES A2 and B2) will result in remarkable changes of vegetation distributions in China (Weng and Zhou 2006): obvious northward shifts of the boreal, temperate deciduous, and evergreen and tropical forests; a large expansion of tropical dry forest/savanna; and reduction of tundra on the Tibetan Plateau. The regions that vegetation types will be changed by climate change take up about 39–49 % of the total terrestrial area of China. These regions are mainly situated on a belt from the northeast to the southwest of China, which lies between the eastern forest zone and the northwestern steppe zone. On the southwest of China, especially in Yunnan province, west of Sichuan province and Hengduan Mountains, the broad-leaved evergreen forest will be replaced by the tropical dry forest/savanna or xerophytic wood/scrub. The western portions of the broad-leaved evergreen forest zone would meet drier climate, and the broad-leaved evergreen forest would reduce from these areas. Most areas of the northern China, where the potential vegetation is temperate deciduous forest under current climate, will be taken by warm grass/shrub and broad-leaved evergreen forest. The Tibetan Plateau, where the biome is predicted to be tundra under current climate, will be covered by conifer forest and cool grass/shrub in its eastern and southwestern portions, respectively. In central China, the temperate deciduous forest will be replaced by the broad-leaved evergreen forest. Northern China, the Tibetan Plateau, and southwestern China (mainly Hengduan Mountains in Yunnan province and west of Sichuan province) are the vulnerable regions sensitive to climate changes.

Under the current climate, NPP of vegetation along the Northeast China Transect (NECT) decreases from east to west. Under the SRES A2 and B2 climate scenarios, the average NPP of each biome along the NECT increased due to the changes of vegetation distribution and climate factors; however, the minimum and maximum NPP exhibited different changes. A new vegetation type, temperate deciduous forest, displayed the largest NPP, ranging from 393 to 629 g/(m^2•a). At the same time, the transect production increased by 4.61 % under the B2 scenario, while it increased by 12.2 % under the A2 scenario. Under the B2 climate scenario, the response of total transect vegetation NPP resulted from the change in vegetation structure and represented −2.85 % of total change in NPP, but the NPP response that resulted from the change in climate factors was 7.69 %. Under the A2 climate scenario, the changes in vegetation NPP are more dramatic than under the B2 scenario. The response of NPP results from the change in vegetation structure and represents −7.4 % of total change in NPP, but the NPP response that resulted from the change in climate factors is 19.56 %.

5.3.3 Impact on Regional Water Resources

Based on the climate change scenarios under SRES A1, A2, B1, and B2, the possible change of runoff in the future is estimated by the variable infiltration capacity (VIC) macroscale hydrological model. Results show that the runoff depth will decrease apparently in Ningxia, Jilin, and Hainan, decrease slightly in Shanxi, and remains the same in Sichuan, while the increase in varying degrees can be found in other provinces. Largest increase in the south is found in Fujian, while in the north is found in Xinjiang. Under A1 and A2 scenarios, the annual mean depth of runoff in the next hundred years over China will increase by 9.7 % (+52 mm) and 9 % (+46 mm), respectively, compared to the present day while under B1 and B2, the runoff depth increase will be less, by 7.5 % (+40 mm) and 8 % (+46 mm), respectively.

For the seasonal distribution, the runoff depth over the northern regions will increase from March to June, and decrease from July to December; over south regions, it will increase in spring and summer and decrease in autumn and winter.

The future significant warming in the northeast and north China together with the decrease of precipitation and runoff depth during summer, especially in the northeast, may lead to more hot and dry days, resulting the tendency of a warm and dry climate in the areas. In northwest, the region of southwestern Xinjiang (Tarim river basin), the precipitation increase in winter and spring and the runoff depth in spring and summer show a possible wetter trend. The summer precipitation and runoff depth will increase in the southern regions of southern part of east, central, south, and southwest China. This is more significant in the southern part of east China, may lead to more flood events over the area. Finally, the winter temperatures over the southern regions will increase more remarkably, while the precipitation and runoff depth will reduce there, especially in Huanan, indicating more frequently winter droughts over the areas. Overall, the present shortage of water resources in the Haihe and Yellow River basins, and the flooding in Zhejiang and Fujian provinces, as well as in the Yangtze and Pearl River basins are not seemingly to be improved from the perspective of climate change.

5.3.4 Impact on Other Aspects of the Region

Climate change results to the continuous sea-level rise. Meanwhile, human activities like the over-extracting of groundwater in the coastal areas lead to the surface settlement. These together result a larger relative sea-level rise, further threatening the security of the coastal areas. The relative sea-level rise may lead to land submerge, increased storm surges, and severe coastal erosions.

The direct impact of the relative sea-level rising is the submerge of large amounts of lands along the coastal regions. The area with elevation less than 5 m along coastal regions is 14.39×10^4 km^2, about 11.3 % of the coverage in the 11 coastal provinces, municipality, and autonomous areas, which also accounts for 1.5 % of the nation's land area. Sea-level rising will inundate large tracts of fertile land and leads to the deterioration of ecological environment in the coastal areas. In addition, it may lead to changes in the structure of local industries and even causes the migration to the further inland areas, finally resulting a series of social and economic problems. It is estimated that the fragile area of the Pearl River Delta, Yangtze River Delta, and the Yellow River Delta and large cities in these regions will be greatly affected in 2050 under the present-day existing storm and flood protection facilities. In the Pearl River Delta region, with the sea-level rise of 50 cm, return period of the storm surge affecting the coast close to Guangzhou will change from the present day 50 years to 10 years. In the other coast areas (e.g., Lantern Hill in Zhongshan and Sishengwei in Dongguan) the return period of 100 may decrease to 10.

The rising sea levels lead to the increase of water depth and tidal range, thus reinforcing the waves and tidal currents. According to the investigations, a doubled water depth could strengthen the wave intensity by 5.6 times. The tide rise of 1 cm will lead to the tidal range increase to 0.34–0.69 cm. The result of the interaction between the two is bound to exacerbate coastal erosion, narrow the high marsh, and thicken the sediments. Further more, the sea-level rise will cause intrusion of brine, resulting the salinization of fresh water both from surface and underground. Besides the shortage of freshwater resources, it will also worsen the condition of land salinization along the coast areas.

References

Dong, W. J., Ren, F. M., Huang, J. B., & Guo, Y. (2012). The atlas of climate change: Based on SEAP-CMIP5. Springer, ISBN 978-3-642-31772-9.

Gao, X. J., Shi, Y., Song, R. Y., Giorgi, F., Wang, Y. G., & Zhang, D. F. (2008). Reduction of future monsoon precipitation over China: Comparison between a high resolution RCM simulation and the driving GCM. *Meteorology and Atmospheric Physics, 100*, 73–86. doi:10.1007/s00703-008-0296-5.

Gao, X. J., Xu, Y., Zhao, Z. C., Pal, J. S., & Giorgi, F. (2006). On the role of resolution and topography in the simulation of East Asia precipitation. *Theoretical and Applied Climatology, 86*, 173–185. doi:10.1007/s00704-005-0214-4.

IPCC. (2013). In Stocker, T. F., Qin, D., Plattner, G.-K., Tignor, M., Allen, S. K., Boschung, J., Nauels, A., Xia, Y., Bex, V., & Midgley, P. M. (eds.), *Climate change 2013: The physical science basis*. Contribution of working group I to the fifth assessment report of the intergovernmental panel on climate change (1535 pp). Cambridge, United Kingdom and New York, NY, USA: Cambridge University Press.

Weng, E. S., & Zhou, G. S. (2006). Modeling distribution changes of vegetation in China under future climate change. *Environmental Modeling and Assessment, 11*, 45–58.

Wu, T. W., Yu, R. C., Zhang, F., Wang, Z. Z., Dong, M., Wang, L. N., Jin, X., Chen, D. L., & Li, L. (2010). The Beijing Climate Center atmospheric general circulation model: Description and its performance for the present-day climate. *Climate Dynamics, 34*, 123–147. doi:10.1007/s00382-008-0487-2.

Xiong, W., Ju, H., & Xu, Y. L. (2006). Regional simulation of wheat yield in China under the climatic change conditions. *Chinese Journal of Eco-Agriculture, 14*(2), 164–167.

Xu, Y., Gao, X. J., & Giorgi, F. (2010). Upgrades to the REA method for producing probabilistic climate change predictions. *Climate Research, 41*, 61–81. doi:10.3354/cr00835.

Yang, X. G., Liu, Z. J., & Chen, F. (2011). The possible effects of global warming on cropping systems in China VI. Possible effects of future climate change on northern limits of cropping system in China. *Scientia Agricultura Sinica*, *44*(8), 1562–1570.

Yang, S. B., Shen, S. H., Zhao, X. Y., Zhao, Y. X., Xu, Y. L., Wang, Z. Y., Liu, J., & Zhang, W. W. (2010). Impacts of climate changes on rice production in the middle and lower reaches of the Yangtze River. *Acta Agronomica Sinica, 36*(9), 1519–1528.

Zhou, T. J., Wu, B., Wen, X. Y., Li, L. J., & Wang, B. (2008). A fast version of LASG/IAP climate system model and its 1000-year control integration. *Advances in Atmospheric Sciences, 25*(4), 655–672.

Chapter 6
Climate Change Mitigation and Adaptation: Technology and Policy Options

Erda Lin, Kejun Jiang, Xiulian Hu, Juncheng Zuo, Maosong Li and Hui Ju

Abstract China has adopted a wide range of measures in the energy sector including energy conservation, renewable energy, and nuclear energy development, as well as in the field of climate change adaptation. These practical actions enabled China to achieve considerable progress and development in climate change mitigation and adaptation, which also provides strong support for China's transition to a low-carbon economy. Under the scenario of global temperature rising more than 3 ° C in the future, implementing adaptation actions in China will require additional capital inputs, new policy guidance, and strengthened research and development of new technologies, in order to offset the negative impacts of climate change. By reinforcing policies and promoting technological advancement, China may expect its CO_2 emissions to peak before 2030, or even 2025, to contribute to meeting the target of limiting global warming to 2 °C.

Keywords Climate change · Greenhouse gas mitigation · Adaptation action · Mitigation potential · Mitigation technology

E. Lin (✉) · H. Ju
Agro-Environment and Sustainable Development Institute, Chinese Academy of Agricultural Sciences, Beijing 100081, China
e-mail: lined@ami.ac.cn

K. Jiang · X. Hu
Energy Research Institute, National Development and Reform Commission, Beijing 100038, China

J. Zuo
Key laboratory of Coastal Disaster and Defense, Ministry of Education, Hohai University, Nanjing 210098, China

M. Li
Institute of Agricultural Resources and Regional Planning, Chinese Academy of Agricultural Sciences, Beijing 100081, China

D. Qin et al. (eds.), *Climate and Environmental Change in China: 1951–2012*, Springer Environmental Science and Engineering,
DOI 10.1007/978-3-662-48482-1_6

6.1 Adaptation and Mitigation Policies and Actions in Recent Years

6.1.1 Adaptation Actions

In recent years, climate change adaptation actions have been widely strengthened in the fields of agriculture, water resources, biodiversity, coastal zones, and human health. China has conducted scientific research related to climate change and impacts assessment of climate change, perfect climate-related laws, regulations, and policies and enhanced the ability to adapt to climate change in key sectors. Based on the local circumstances, the state has strengthened construction of agricultural infrastructure and construction of farmland water conservancy, renovated the supporting facilities and large-scale pumping stations in major irrigation zones, enlarged the area of agricultural irrigation, improved farmland irrigation and drainage efficiency and capability, and promoted dry farming and water-saving technologies. China has stepped up the construction of projects to control floods on major rivers as well as systems to control floods caused by mountain torrents, and meantime restricted groundwater extraction. The country has continued to promote key forest projects, including protecting natural forests, restoring farmland to woodland, preserving natural wildlife habitats, and conserving wetland. China has pushed forwarded sustainable development and management of forest, intensified its efforts in ecological water and soil conservation, and restored eco-fragile areas and the functions of ecosystems. China has upgraded protection standards for coastal cities and major infrastructure, and actively carried out the cultivation and transplantation of mangrove forests, transplantation and protection of coral reefs, restoration of coastal wetlands and beach, and other marine ecological restoration demonstration projects. Moreover, the country has initiated the observation and early-warning system for storm surge, sea wave, tsunami, sea ice, and other marine disasters, as well as the monitoring, investigation, and assessment of sea level rise, coastal erosion, seawater intrusion, and soil salinization (Cai et al. 2008). A series of adaptation measures have also been implemented in the field of human health and human habitat. China has established and improved the public health disease surveillance system, the public health emergencies responding mechanism, the disease prevention and control system, and the health supervision and law enforcement system.

The next few years will see the continuation of adjustments of agricultural structure and cropping system and the construction of benign agroecology cycle system, taking into account the multi-level use of materials and energy and waste recycling and reuse. Climate change will be integrated into water resources assessment and planning: Based on the carrying capacity of water resources, China will reinforce the integrated management of water resources, enhance the water allocation capacity, construct a sound information system for water resource management, and raise design standards for water conservancy projects. Besides, China will step up its effort in the following areas: afforestation, capacity for forest fire control, pest, and disease prevention and control; adjust the way and timing of pasture grazing, and enlarge the area of irrigated grassland and artificial pastures; strengthen wetland ecological protection and pollution control in rivers and lakes; protect biological resources in deserts, combat desertification, and reinforce situ-protection of local species; and build breeding bases for endangered species, strengthen the breeding of rare and endangered species, and increase the natural adaptation capacity of various ecosystems. In coastal areas, China will improve the design standards for tide-resistant facilities, heighten and consolidate existing facilities (Du and Shi 1993); implement coastal protection projects, develop coastal forest shelterbelts; formulate relevant standards for land subsidence observation as well as for sewage recharge; develop standards for groundwater level observation, groundwater quality monitoring; draw up standards for drinking water source and effluent discharge; gradually increase the investment in construction of sea and river banks; optimize the arrangement and aquifer for groundwater exploration, and fully take use of shallow groundwater; accelerate salt water conversion and utilization. It will also reinforce controls on tropical diseases that have been previously neglected, and set up and perfect a system to monitor and issue early warning about the effects of climate change on human health.

6.1.2 Mitigation Actions

During the 11th Five-Year Plan period, China has advanced its policies and actions for mitigating climate change and implemented a range of policies and measures relating to the transformation of economic growth patterns, energy conservation and efficiency enhancement, optimization of the energy mix, and afforestation. These actions include adjusting the economic structure, accelerating the optimization, upgrading the industrial structure, developing renewable energy, promoting circular economy, reducing GHG emissions, controlling GHG emissions in the agriculture sector and in rural areas, enhancing the carbon sequestration capacity of ecosystems, and intensifying scientific research to address climate change scientifically.

In the field of energy, China will continue to vigorously advance the construction of large-scale hydropower and wind power plants, popularize the photovoltaic roof system, build large-scale demonstrative grid-connected photovoltaic power plants, and actively develop power generation through biomass combustion, co-firing, and gasification. Energy-saving policies and measures will be reinforced in transportation, building, agriculture and land use, forestry, animal husbandry, and other sectors.

In the field of transportation, China will continue to implement fuel economy standards, draw up policies and regulations to control vehicle air pollution, set up the labeling mechanism to increase public awareness, and formulate incentive as well as preferential financial and tax policies to popularize energy saving and new energy vehicles. Regarding the building sector, the state will intensify its energy-efficiency policies, complete the market mechanism, improve the technical level, and boost the application of renewable energy technologies and a range of emerging technologies with higher energy-saving performance. Agricultural GHG emissions will be controlled mainly through increasing carbon storage in "sinks"—sequestration of atmospheric CO_2 by soil, and reducing carbon emissions from "sources"—curbing CH_4 and N_2O emissions from agricultural activities. As forests play an important role in climate change mitigation, China will multiply its efforts to protect and develop forests and intensify afforestation and improve forest management. As to the animal husbandry sector, China will continue to promote grassland-based livestock production, provide high-quality forage to livestock, improve the livestock feed conversion ratio, and reduce GHG emissions from ruminants (Huang 2006).

China will continue to build the policy system for sustainable development and transform economic development patterns. At the same time, the country will adjust the spatial pattern of land development, implement the planning of development priority zones, optimize the economy layout, promote regional coordinated development, and give special attention to the preservation and construction of natural reserves, eco-fragile areas, and wetlands. Moreover, China will adhere to the family planning policy, reinforce the poverty alleviation policy, intensify energy management, and integrate energy consumption targets into the annual assessment and evaluation system of local economic and social development in the attempt to reduce energy use. In addition, it will reinforce the implementation of the natural forest logging ban, the restoration of farmland to forest and grassland, afforestation, grassland construction and management, rehabilitation of degraded land, soil erosion control, and wetland protection, and therefore enhancing the sinks' capacity to store carbon.

6.2 Existing Adaptation and Mitigation Technologies

6.2.1 Adaptation Technologies

In the context of addressing climate change, China has actively carried out research on climate change adaptation and mitigation, laid emphasis on capacity building to adapt to climate change, developed a series of effective adaptation technologies, and achieved remarkable effects.

China has taken effective actions to promote technologies for disaster prevention and mitigation and climate adaptation in the agriculture and forestry sectors, which have effectively improved the two sectors' ability to adapt to climate change. Major actions include the following:

Raise the ability to prevent and mitigate biological disasters and to adapt to climate change: China has developed monitoring and warning technologies for major agrobiological disasters, detection technologies of plant pathogens by molecular methods, and other key technologies. Consequently, a new technology system has been built for biological disaster detection, monitoring, and warning, and the prediction accuracy has been improved remarkably with the average accuracy higher than 80 %. Research has also been carried out to develop agricultural pests monitoring systems and light trapping and controlling technology systems. Besides, meteorological data have been collected to enable the remote real-time pest and disease monitoring and information sharing among stations for pest and disease monitoring and reporting throughout the nation. In addition, modern information technologies have been harnessed to build an integrated rice pests GIS monitoring and warning system, with over 95 % accuracy for short-term predictions. In the field of forests, service-oriented monitoring, warning, and management systems have been developed for forestry pest control, which enabled the visualized multi-level management of forestry pests. Moreover, the open and dynamic integration of disaster monitoring services and model-based prediction services will provide a useful platform for new technology application and diffusion to monitor and manage forest pests' breakout.

Agro-meteorological disaster prevention and mitigation technologies to adapt to climate change: Disaster prevention and mitigation techniques and measures include selection of disaster-resistant varieties, application of efficient and suitable cultivation modes, timely sowing, establishment and improvement of nursery centers, scientific fertilization, reasonable irrigation and drainage, covering protection, chemical control, cost-saving and yield-increasing cultivation techniques. Based on the integration and amelioration of these techniques and practices, key disaster prevention technology systems for different kind of crops have been instituted. As a result, the country's disaster prevention capacity in crop production has been consistently improved.

Improve crop varieties, adjust the distribution of crops, and optimize the cropping system to adapt to climate change: According to local conditions, highly adaptable varieties have been introduced, selected, and bred to achieve the goal of disaster resilience, high and stable yield, and high quality. In the northern arid and semiarid regions, more than 80 % of wheat, maize, and cereals (millet, Panicum miliaceum) are able to withstand drought and save water. Developing cold- and frost-resistant varieties and improving crops' ability to withstand unexpected cold, and chilling and frost disasters in low-temperature winters will ensure stable yields in a region. The formulation of climatic risk zoning for sugarcane, cassava, banana, litchi, longan, and other subtropical crops optimized the distribution of crops and reduced chilling and frost losses by more than 5 % compared with the current distribution pattern. To promote the steady development of agriculture, adopt new farming practices, implement conservation tillage, dry farming, and efficient fertilization. China has conducted research on and developed appropriate conservation tillage techniques to accommodate regional climate and resource situations and cropping systems, such as deep loosening and wide-narrow row alternate planting of maize in the northeast, straw mulching and stubble retention tillage techniques in hilly and gully regions in the northwest, key conservation tillage techniques for sandy fields, multi-ground covering techniques, and retention of maize, and wheat stubbles will be greatly beneficial to soil and water conservation in these regions. In dry sandy dam farmlands of North Heibei Province, growing pumpkins with film-mulching and ridging achieved good drought-resistant and yield increase effects.

Water resources utilization and water-saving irrigation techniques to adapt to climate change include following actions: 1) Continuously reinforce the network construction of hydrological stations and groundwater monitoring stations within administrative borders based on hydrology and water quality monitoring and warning technologies; 2) Raise the accuracy and timeliness of forecasts by strengthening the construction of storm flood predictions, forecasting, and warning facilities; 3) Increase the water use efficiency in dry areas, like develop oasis agricultural water-saving technologies, drip irrigation under mulch and water–salt–fertilizer coupling and regulating to cotton fields, tiny irrigation to mature fruit trees, and brackish water irrigation on cotton. In the meantime, we need to vigorously enforce the protection of forest lands, forests and wild animals and plants; continue to advance the protection of natural forests, wildlife and wetland; and strengthen the restoration and rehabilitation of ecologically fragile areas and ecosystems, including restoring farmland to forest land, the Three-North shelter forest project, controlling the sources of Beijing and Tianjin sandstorms, the Yangtze River Basin shelter forests project, building the coastal and farmland shelterbelt system. Forest fire and pest control abilities have also been significantly improved.

6.2.2 *Mitigation Technologies*

Since GHG emissions are closely related to energy consumption, mitigation technologies in the energy sector play an important role in reducing GHG emissions.

In the Chinese energy supply sector, the following technologies have been deployed: advanced and efficient (ultra-) supercritical coal-fired power generation and natural gas power generation; clean and efficient integrated gasification combined cycle (IGCC) technology; advanced nuclear power, hydropower, solar, wind, geothermal, biomass energies, and other new energies; and applying carbon capture and sequestration (CCS) technologies to coal-fired power plants as early as possible can stipulate 16 million tons of emission reductions of by 2020 and substantially reduce the dependence on polluting coals.

In terms of mitigation potentials, the nuclear power and hydropower have the largest abatement potentials, individually accounting for nearly one-third of the total mitigation potentials of the energy supply sector. They are followed by the wind power, whose mitigation potentials can reach 270 million tons of CO_2 by 2020, making up 17 % of the sectoral total. In addition, (ultra-) supercritical technologies are responsible for 8 % of the overall mitigation potentials, amounting to 130 million tons, while the role that IGCC and CCS technologies can play will be very limited in the near future (Zhang et al. 2011).

In terms of abatement costs, except for (ultra-) supercritical technologies and hydropower technologies with negative costs, abatement costs of the vast majority of these technologies range between 30 and 100 $ /tCO_2, among which the nuclear power, wind power, and IGCC are relatively low-cost abatement technologies and should be developed in priority in the near future. This indicates that the application and deployment of low-carbon energy technologies in China entails important economic costs.

Supercritical (Ultra-) power generation technologies are already technologically mature and hold significant efficiency advantages and cost competitiveness compared with the conventional subcritical power generation technology. Their average efficiency can reach 39–45 %, and costs of electricity generation are about 13–15 % lower than conventional technology. IGCC is one of the most clean and efficient coal-fired power generation technologies, and currently its system efficiency is 40–47 % (the highest efficiency is expected to be up to 50 %). IGCC is applicable to all carbon-containing raw materials including coal, oil, coke, biomass, and municipal solid waste and has various advantages such as generating less pollutants and emissions (including sulfur compounds, nitrogen oxides), high potential for efficiency improvement, and enabling effective control of CO_2 emissions (Zhang et al. 2011).

From the economic point of view, nuclear energy has the features of high initial investment, low running cost, and long design life (40–60 years); for example, second- and third-generation nuclear power plants require an initial investment at about 15,000–18,000 yuan/kW, and their investment payback periods stand at 10–15 years. As the cost of nuclear power generation is primarily affected by the load factor, the higher the running rate of nuclear power plants, the lower the cost of power generation. As a result, it is appropriate to use the nuclear as the energy source for base load power. By 2020, the installed nuclear power capacity in China may reach 70–80 GW.

Hydropower is a mature renewable energy power generation technology and has already been widely deployed around the world. Hydropower is also a major renewable energy in China and tops the list of lowest cost power generation technologies, and in general it has a long operating lifetime.

There are several ways of turning biomass into electricity, including direct combustion, co-firing, and gasification, and these are the most technically mature technologies with the largest development scale among modern biomass energy use technologies. From 2020 onwards, in addition to those mitigation technologies discussed above, solar thermal technology, smart grid, low-cost CCS technology, fourth-generation nuclear energy, and other technologies shall also play an important role. In particular, coal-based integrated gasification fuel cell (IGCC) power plants will be built, which directly turn heat and methane chemical energy into electricity, using hydrogasifier integrated with fuel cells (FC). The overall energy conversion efficiency of the IGFC process is about 1/2 to 3/4 higher than the IGCC system and enables substantial savings in coal consumption and reductions of CO_2 emissions. Its prospects for development should by no means be underestimated.

CO_2 emission mitigation technologies and practices have been in constant advancement, and concentrate in the industry, transportation, building, and other energy end-use sectors. Research results show that these sectors will be mainly responsible for curbing CO_2 emission growth in China both currently and in the future. Continuous promotion of advanced, efficient, low-carbon emission, and cost-effective CCS technologies shall allow for the realization of around 2.2 billion tons of CO_2 emission reduction potentials by 2020. The industry, transportation, and building sectors constitute 45, 30, and 25 % of the total mitigation potentials, respectively. By 2050, the mitigation potentials from energy end-use sectors will stand at approximately 4.8 billion tons, among which 37, 35, and 28 % come from the industry, transportation, and building sectors, respectively (Hu et al. 2007).

The industry sector is the priority and focus area to realize the technical mitigation potentials of China's energy end-use sectors. Before 2020, major abatement technologies in the industry sector include those improving energy efficiency, new processes and new technologies, by-products and waste recycling and reuse technologies, substitutes of raw materials and fuels, and CCS. Analysis on the marginal abatement cost of some industry mitigation technologies implies that over 40 % of these technologies will be cost-effective to 2020. After 2020, besides those aforementioned, a large extent of other low-carbon technologies will be developed and applied, for example, low-carbon steel-making technology, top gas recycling blast furnace technology (TGRBF), advanced direct-reduced process (ULCORED), new melting reduction process (HIsarna), electrolysis of iron ore in the iron and steel industry, eco-cement production technology in the cement industry, and CCS in the cement and chemical industries.

Transportation is an energy end-use sector with the fastest growing technical potential for emission reductions. Before 2030, road transportation will provide as high as 85 % of the overall mitigation potential in this sector, but declining to about 60 % in 2050. Major mitigation technologies and measures in the transportation sector include those improving the economic efficiency of fuels; promoting efficient gasoline internal combustion engines, alternative fuels, and efficient diesel internal combustion engines; spreading the application of electric vehicles, super-efficient diesel, hybrid fuels, hydrogen power, fuel cell, hybrid power, advanced diesel vehicles; and developing and applying alternative materials and energies (such as light materials), renewable energies, and alternative ways to replace transportation. Among them, mitigation costs of efficient gasoline internal combustion engine, alternative fuels, and efficient diesel combustion engine technologies stand below 150 yuan /tCO_2, while the plug-in hybrid vehicles, hybrid vehicles, and pure electric vehicles can only generate emission reductions at a cost higher than 600 yuan/tCO_2 (Hu et al. 2011).

The building sector is another potential area being crucial to China realizing technical mitigation potentials. Because emission sources as well as relevant abatement technologies are diverse in this sector, energy-saving lighting technology is attached the highest relative mitigation potential, which emits 70–90 % less GHG compared with traditional lighting system. In addition, use of renewable energies and energy-saving technologies for new buildings and building envelopes enables high emission reductions. In terms of absolute mitigation potentials, application of energy-saving technologies for building envelopes, heating systems, and lighting systems will greatly reduce emissions. Design of new energy-efficient buildings is also expected to generate important emission reductions, which essentially fall upon new energy-efficient buildings in the short term. In contrast, passive house design and green buildings will only have a role to play with a longer time horizon. Due to combined effects of building type, size, geographic location, climate, and policies, great uncertainties exist in assessing costs of different abatement technologies in the building sector. If only intrinsic factors are taken into account, the deployment of cost-effective technologies is able to deliver approximately 65 % of the total mitigation potential from the building sector.

Biological carbon sequestration technologies: Besides mitigation technologies in the energy and industry sectors, carbon sequestration and non-CO_2 GHG emission reduction technologies have also been developed in agriculture and forestry, including the following:

Carbon sequestration and emission reduction technologies in the farmland ecosystem: Straw return is a major energy conservation and mitigation practice in agriculture and has long been advocated by various departments. Use of nitrification inhibitors can prohibit N_2O emissions from dry farmlands; improved organic fertilizing and tilling method for turfy soil can increase soil carbon sequestration. Conservation tillage practices such as straw return and water–fertilizer coupling and regulating are conducive to reducing GHG emission from and storing carbon in paddy fields. Applying biochars in either the rice-growing or wheat-growing season improves the activity of soil microbial flora and enzymes and increases soil carbon sequestration.

Carbon sequestration and emission reduction technologies in the forest ecosystem: Research has been conducted to find variation rules of carbon intensity and carbon storage in forests with different provenances or in distinct families, in the purpose to select tree species, provenances, or families with high carbon sequestration capacity. The carbon sequestration capacity has been raised in vulnerable areas through afforestation and forest management and protection.

Wetland protection, carbon sequestration, and emission reduction technologies: These technologies prevent wetland degradation and minimize the loss of soil organic carbon.

Carbon sequestration technologies in the grassland ecosystem: Grassland enclosure can significantly improve aboveground live plants, litter/underground live roots, and soil carbon density and soil carbon storage in grassland ecosystems including the Loess Plateau.

Carbon sequestration and emission reduction technologies in the marine ecosystem: China has carried out demonstrative carbon storage projects, including bottom sowing culture in typical beaches in North China and shallow shellfish and algae culture in shallow raft areas.

6.3 Adaptation and Mitigation Outlook in China

6.3.1 Adaptation Outlook

1. Arduous task of adaptation

As a developing country with a large population, relatively low level of economic development, a complex climate, and a fragile co-environment, China is vulnerable to the adverse effects of climate change, which has brought substantial threats to the natural ecosystems as well as the economic and social development of the country. The multiple pressures of developing the economy, eliminating poverty, and protecting the environment constitute difficulties for China in its efforts to cope with climate change, since the country is undergoing rapid economic development. Therefore, adapting to climate change has emerged as one of China's practical requirements and priority tasks (Sun et al. 2013).

Climate change adaptation is not only a natural scientific issue, but also an issue in social science having policy, environmental, and economic implications. The complexity of climate change adaptation and the current status of social and economic development imply multiple difficulties for China to adapt to climate change. In the first place, though China is in the stage of industrialization, it no longer enjoys the resources and environment that favored the rise of the developed countries. China has a complex climate, a fragile eco-environment, extremely uneven economic development among regions, and weak capacity to adapt to climate change with significant regional discrepancies. Secondly, as the basic research relating to climate change adaptation is not advanced and China is experiencing rapid urbanization, marketization, and industrialization, development continues to be the dominant issue in many sectors; climate change issue has not emerged as a priority, and public awareness still needs to be raised. Furthermore, as China is in a period of high-speed economic development, the adaptation funding channel is yet to be perfected; however, resource and environmental capacity cannot afford delayed action on adaptation, therefore resulting in enormous challenges for China to adapt to climate change.

2. Improve scientific understanding of adaptation

The purpose of climate change adaptation is to safeguard people's livelihood, public and private enterprises, assets, communities, infrastructure, and the economy's ability to adapt to climate change. Various climate change-related studies have been actively carried out both domestically and internationally, providing the theoretical basis for tackling climate change. The international trends indicate that research on global change impacts and adaptation will become not only the nucleus in the scientific field over the coming years, but also the focus of attention of the international community.

A handful of scientific problems remain unsolved in China regarding climate change adaptation. The premier task is to analyze the dynamics of elements contributing to climate change and their causes. In essence, climate change affects social, environmental, and economics through variations in the changing rate, intensity, and frequency. Climate change-related research must track alterations in resource and environmental elements and analyze the changing dynamics, assess their change characteristics, identify critical thresholds of climate change impacts on various systems and individuals as well as the regions and fields likely to be affected, understand the relative contribution of natural and human factors to climate change, identify underlying climate risks in different fields under specific natural and socioeconomic conditions, conduct scientific assessments, and set adaptation goals.

Strengthen scientific research to objectively understand the process and behavior of the human society to adapt to climate change. Accurate understanding of the processes and impacts of climate change is the key premise in climate change adaptation decision-making, and adaptation option prioritization. The principle of selecting adaptation measures as well as the causes and consequences of delayed adaptation actions has not been well explored in the adaptation research domain. Adaptation measure prioritization should rely on the accurate cost-effective analysis of measures. In addition, assessment of adaptation measures should not only consider economic benefits, but also take into account the ecological and social benefits as well as political factors, therefore calling for more comprehensive and in-depth assessment of adaptation achievements.

3. Strengthen policy guidance and capacity building

The impacts of climate change on resources and eco-environment systems cannot be ignored in China, especially in the sensitive fields such as agriculture, ecology, energy, transportation, and human health. Although the certainty of impact understanding has yet to be improved, some recent studies have reached clear conclusions: Based on the current level of understanding, we should elect "no-regret options and measures" that will promote both climate change adaptation and social development. In addition, implementation issues should be incorporated into national socioeconomic development plans, in order to prepare beforehand, avoid disadvantages and pursue advantages, and ensure sustainable and healthy socioeconomic development in China.

As a developing country, China should strengthen capacity building on climate change in the guidance of scientific theories. Climate change adaptation is a long process rather than a one-time action, as changing climate requires the society and individuals to timely adjust their behaviors. China should raise public awareness on climate change in order to adapt to climate change and extreme weather events, in which process supportive and encouraging policies will play a positive and guiding role. Measures in this regard include the following: raise the awareness of policy makers at all levels on the issues of climate change and gradually improve the capacity for action on climate change, establish incentive mechanisms to trigger extensive public participation, give full play to the role of public supervision, actively give play to the initiative of social communities and non-governmental organizations, and encourage all circles of the society to take climate change adaptation actions.

4. Implement adaptation actions

Adaptation actions further enhance agriculture's ability to adapt to climate change through strengthening farmland infrastructure, adjusting cropping systems, selecting and breeding stress-resistant varieties, and developing biotechnologies and other adaptive countermeasures; protect typical forest ecosystems and key species of wild fauna and flora, restore desertified land, and effectively improve the adaptability of forest ecosystems through strengthening the natural forest conservation and nature reserve management, continuously implementing key ecological restoration programmes, and establishing key ecological protection areas; reduce the vulnerability of water resources to climate change through rational exploitation and optimized allocation of water resources, popularization of water-saving measures, upgrading the anti-drought standards of farmlands and other measures; remarkably raise the capability of costal zones to resist marine disasters through scientific monitoring of sea level change, reinforcing related regulation, and advancing the construction of coastal shelterbelt system.

Reinforce the implementation of adaptation actions in key areas and regions: A lot of adaptation attempts in the agriculture, water resource, natural ecosystem, and other sectors have proven to be successful, which have important guiding significance to the full implementation of adaptation actions in key areas and regions in China (Ju et al. 2011). In the future, provincial and sectoral climate change adaptation programmes will be widely initiated. In the meantime, China will continue to implement the National Climate Change Programs, including restoring farmland to forest and grassland, improving water use efficiency in irrigated agriculture, advancing grassland improvement, desertification treatment, and protecting natural forests.

5. Adaptation actions require additional capital inputs

Integrated research on the impacts of climate change on agriculture, water resources, nature ecological systems, and coastal zones shows that if the global average (surface) temperature rise is limited to no more than 2 °C by the 2050s compared to pre-industrial levels, climate change will generate both positive and negative impacts in China; existing adaptation technologies will be able to offset the adverse impacts and highlight the potential of positive impacts; investment put into adaptation shall bring positive return in general, although adaptive abilities and adaptation degrees will vary among regions. If the temperature rise is below 3 °C, the advantage-promoting capacity and disadvantage-avoiding capacity of existing adaptation technologies will gradually shrink, and it will be necessary to develop new sectoral and regional adaptation measures through technology R&D, policy planning, and capacity building, and to increase investment to fully tap their technical superiority in a bid to offset the adverse impacts. A global average temperature rise of 4 °C will induce substantial additional costs for China's sectors and regions to adapt to climate change as sole application of existing technologies will be insufficient to offset the adverse impacts of climate change; what is worse, it would be difficult for some regions and sectors to effectively control or compensate the negative impacts of climate change even substantial fund is dedicated to adaptation, which may even trigger unfavorable results. However, it should be noted that current studies on climate scenarios, both in China and abroad, have not reached a consensus on key factors, driving forces, action mechanisms, and feedback mechanisms of the interactions between different systems. The estimated impacts of climate change are far from accurate and there are still many limiting factors unknown which influence the assessment of climate change impacts on ecosystems and social-economic sectors under different warming scenarios, calling for more in-depth research on adaptation ability and adaptation inputs.

Fact box 6.1: Assessment of future losses associated with climate change in China.

China is one of the countries in the world most adversely affected by meteorological disasters: typhoon, torrential rain (snow), thunder and lightning, drought, strong wind, hail, fog, haze, sandstorm, heat waves, low-temperature damage, and other disasters often happen. Every year, over 70 % of the territory, more than 50 % of the population and about 80 % of industrial and agricultural regions and cities are subject to, to varying extent, meteorological disasters. Meteorological disaster-induced landslides, mudslides, and flash floods, as well as agricultural disasters, marine disasters, biological disasters forest, and grassland fires are quite serious, exerting significant impacts on economic and social development, people's lives and ecological environments and causing considerable economic losses, equivalent to 1–3 % of the national's GDP (Liu and Yan 2011).

In the context of global climate change, increased risks of natural disasters pose more challenges for disaster prevention and mitigation. The risk of drought, flood, typhoon, low temperature, ice and snow, heat wave, sandstorm, pest, and disease disasters is on the rise, while avalanche, landslide, mudslide, flash floods, and other disasters maintained high occurrence frequency. New changes appear in the spatial and temporal distributions of natural disasters, the amount of losses, and the intensity and scope of impacts, and consequently the unexpectedness, irregularity, and unpredictability of natural disasters are becoming more prominent. Along with the markedly accelerated process of industrialization and urbanization and the increase in urban population density, the loading capacity of infrastructure continues to rise and impacts of natural disasters on cities are getting worse; in vast rural areas, especially in central and Western regions, where the economic and social development is relatively backward and the disaster prevention system if far from robust, rural resident's abilities to resist disasters are relatively weak.

The risk of secondary and derivative disasters triggered by natural disasters remains high. Currently, there are two issues requiring our high attention: first, high-risk cities and undefended rural areas coexist in China. Since the reform and opening up, China has been experiencing rapid urbanization and urban modernization with an average annual increase of 16 cities and 1.4 % of urban population. China's urbanization rate has reached 53.7 %, and the number of metropolitans hosting millions of people increased to 140 (National Plan on New-type Urbanization 2014). However, disaster prevention and mitigation infrastructure has not evolved at the same pace and is still weak in China with significant geographical discrepancies among eastern, central, and Western regions, and between urban and rural areas. Urban disasters begin to show the characteristics of unexpectedness, complexity, diversity, linksystem, focuses, seriousness, and amplification. The second issue is that natural disasters and various types of emergencies are becoming more closely correlated; they mutually affect each other and are even mutual convertible, leading to frequent secondary, derivative, and coupled events.

6.3.2 Mitigation Outlook

The IPCC Fourth Assessment Report suggests the required emission levels for different groups of stabilization scenarios: under the first category with CO_2-equivalent (CO_2-eq) concentrations between 445 and 490 ppm, global average temperature is predicted to increase by 2.0–2.4 °C and global CO_2 emissions to decline by 50–85 % in 2050 compared to 2000 levels; under the second category with CO_2-eq concentrations between 490 and 535 ppm, global average temperature is predicted to increase by 2.4–2.8 °C and global CO_2 emissions to decline by 30–60 % in 2050 compared to 2000 levels; under the third category with CO_2-eq concentrations between 535 and 590 ppm, global average temperature is predicted to increase by 2.8–3.2 °C and global CO_2 emissions to decline by 30 % to rise by 5 % in 2050 compared to 2000 levels. These three mitigation scenarios are most widely used in current international modeling studies as well as international cooperation and discussions (IPCC 2007).

According to analysis on various burden sharing methods, China is under great pressure to contribute to achieving the global emission reduction targets. In order to limit the global temperature increase to 1.5–2 °C, China's GHG emissions need to peak between 2020 and 2035, with a total amount to be less than 9 Gt CO_2 (Jiang et al. 2013).

Current model studies suggest that China's rapid economic development will lead to significant increase in energy use. Model simulations under a variety of scenarios show that energy demand in China will grow by 50–100 % by 2030 from 2008 and by 85–150 % by 2050. From another point of view, this implies huge potential for energy savings in China. The growth of national energy demand will gradually slow down after 2020—an important shift from the current energy use growth trend, if appropriate energy conservation policies are put in place (e.g., intensified energy conservation). It is also likely that energy demand would fall to 4.8 billion tons of coal equivalent (tce) by 2030 and 5.5 billion tce by 2050.

In the meantime, China's GHG emissions will grow rapidly. Figure 6.1 shows the results of some recently published modeling studies. Important discrepancies can be observed among these results. An investigation on model parameters shows

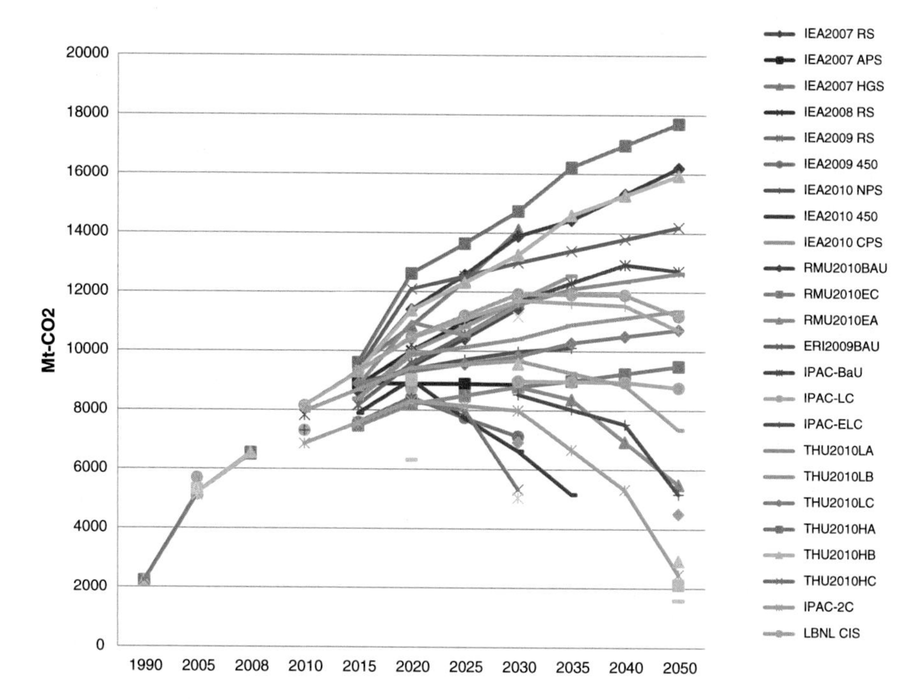

Fig. 6.1 CO_2 emission scenarios from energy activities in China

that all models adopt a baseline emission scenario of more than 11 Gt CO_2 in 2030, which rises to 12–17 Gt in 2050, while emissions may also peak between 2040 and 2050 in some scenarios. Under the scenario with climate change policies being implemented, CO_2 emissions will range between 8.5 and 11 Gt in 2030 and 50 and 80 Gt in 2050. Model studies suggest that limiting global temperature rise to 2 °C would require China's CO_2 emissions to peak at less than 9 Gt before 2025 and to further decline to about 2.5 Gt in 2050. IEA's modeling result under the 450 ppm scenario calls for a peak at 9 Gt in 2020 and a quick decline afterward. CO_2 emission is predicted to peak at 8.4 Gt under the IPAC's 2 °C scenario, and decline afterward, however, at a lower rate than that under the IEA scenario (IEA 2012; Jiang et al. 2013). Despite the rapid increase in CO_2 emissions in China until 2011, the emission reduction feasibility analysis conducted by the IPAC model group shows that the 2 degrees of scenario is still achievable. The rapid technological progress and policy provides strong support for pursing this emission reduction pathway.

In these scenarios, all policy scenarios enable the fulfillment of national CO_2 emission reduction targets, e.g., reducing CO_2 emission intensity of the GDP by 40–45 % in 2020 compared to 2005 levels, while a 54 % decrease in CO_2 intensity can be expected in the lowest emissions scenario.

6.3.2.1 Technology and Policy Implications in Emission Reduction Scenarios

There are some common points relating to policy integration in scenario building. All mitigation scenarios have taken into account the development of renewable energy and nuclear power, energy saving, as well as the application of CCS technology. As a large developing country in a period of economic takeoff, technologies are essential to the energy conservation and efficiency, environment, and climate change. In China, technological progress will play an important role in curbing GHG emissions, and more importantly, most of these technologies have synergies for energy saving and environmental protection in both the short term and the long term.

The full application of low-carbon technologies is a key vehicle to achieve low-carbon development. This argument is also adequately evidenced in the study on China's low-carbon development scenarios. In other words, the development and deployment of some key technologies are crucial for China realizing low-carbon and strengthened low-carbon scenarios in the future. In order to accomplish low-carbon development goals under the low-carbon and strengthened low-carbon scenarios, different technologies will have a role to play in both the energy production sphere and the energy consumption sphere. Before 2030, low-carbon development will be primarily driven by low-carbon technologies in the energy consumption sphere, while will largely fall on technologies in the energy production sphere after 2030.

Nowadays, the major carbon abatement technologies with the largest market development potential and economic benefits in China are those on energy saving. As a result, we should expand their application range in all social sectors, making full use of the favorable opportunity offered by the national energy-saving campaign. In the industry sector, we should seize the opportunity of China's new investment in a large-scale infrastructure, combined with suitable policy measures, to vigorously promote various energy-saving technologies and measures, in the attempt to make China's industry among the most energy-efficient ones on a comparable basis between 2020 and 2030. In the building sector, we should promote energy-efficient buildings in cities, promote a variety of low-cost energy-saving buildings in rural areas, and adopt more stringent energy-efficiency standards, in the attempt to raise building energy efficiency by 30–50 % by 2030 compared with 2010 levels. In the transportation sector, we should step up efforts to promote the public transport as the main means of transportation in large cities by 2030, to popularize energy-efficient cars, and to substantially develop pure electric vehicles.

More incentives and stronger support should be made to stimulate the development of clean energy technologies, including supportive and incentive policies as well as establishing and perfecting market-based mechanisms. A lot of clean energy technologies, such as onshore wind power technology, have already had strong market competitiveness and development potential, and their mitigation effects will be largely catalyzed in the short term by adequate policy supports and favorable market environments. For a number of relatively new technologies (such as hybrid vehicles, and advanced and efficient diesel vehicles) that have gained some market space in developed countries, China should consider to accelerate the introduction of these technologies and enlarge their application range. In addition, the state should also intensify its efforts to encourage the use of some off-grid renewable energy technologies, such as household renewable energy technologies (including household solar thermal technology, household photovoltaic technology, and household wind technologies), to advance their large-scale development before 2030.

Scenario analysis also suggests that China is facing huge challenges to reduce GHG emissions. First, China is in a development stage of accelerated industrialization and urbanization; experience from developed countries has proved that energy consumption and GHG emissions will inevitably grow along with large-scale infrastructure construction in the process of industrialization. Secondly, China's energy resource endowment determines that coal will continue to dominate China's energy mix for a long period in the future, and consequently, energy use in China is relatively more carbon intensive than in other countries. Finally, the current economic and technological strength does not enable China to take on large-scale emission reductions. The extent to which China can contribute to control emissions depends on the medium- to long-term mechanisms set for emission reductions as well as the level of funds and technical support that can be secured to sustain these mechanisms.

Climate change policies can offer a good opportunity for China to transform the economic structure and promote the development of low energy consumption and high value-added industries. Although large-scale emission reductions are not in line with current national conditions, a certain mitigation activities can be undertaken, combined with domestic energy, agricultural development, and land use policies, to help with the transformation of economic development patterns in China.

6.3.2.2 Mitigation Costs and Benefits

As we can see, when assessing mitigation scenarios, most modeling studies followed similar mitigation pathways, however, at varying costs. Modeled costs for a same mitigation technology may even be contradictive between two studies: One study may deem the technology among the high cost ranges while the other concludes low cost for it. Generally speaking, the cost results depend on the modeling method and parameter setting. As technical models focus on the cost of technology, the generated costs are generally low. Macroeconomic models estimate the economic cost of emission reductions, which is relatively high. Abatement costs depend to a large extent on the assumed stabilization level of GHG and the baseline scenario. Estimated abatement costs for the same region may vary substantially due to the differences among modeling methods and relevant assumptions.

Among the models that integrate cost component into scenario assessment, the IPAC model of the Energy Research Institute and the McKinsey scenarios, as well as the PENE model of the Renmin University, are the most representative ones. The IPAC model and the McKinsey scenarios mainly evaluate technical costs of abatement technologies and assume significant decline in abatement costs, therefore reporting relatively low level of technical costs. In addition, there is no considerable rise in costs under the low-carbon scenario compared with the baseline scenario, with even a slight decrease, and the total additional investment accounts for a low proportion of the GDP. The additional investment under the low-carbon scenario in the IPAC model amounts to 1.5 trillion yuan/year, accounting for about 1.2 % of the GDP of the year. McKinsey's research suggests that about 1/3 of the investment will generate economic benefits, another 1/3 induce low to moderate degrees of economic cost, and the last 1/3 imply substantial economic cost. The Remin University conducted abatement costs analysis based on the PENE economic model, using 2005 as the base year, and concluded that GDP will fall 4.6 and 9.9 %, respectively, under the emission control scenario and emission reduction scenario, compared to the baseline levels in 2050. Such economic losses correspond to 1 trillion and 2.3 trillion US dollars, respectively, and account for 46 and 97 % of the GDP in 2005 based on 2005 constant prices. GDP losses per ton CO_2 stand at 158 and 210 dollars. The IPAC model and the McKinsey scenarios do not fully consider other social costs besides technological cost; in contrast, the PENE model pays limited attention to the role of technological progress. As cost estimates of different modeling systems are often inconsistent, their modeling methods and parameter setting should be investigated to draw reasonable comparisons. Such kind of in-depth modeled analysis can provide a cost range and thus permits a better understanding of costs.

6.4 Implementation Approaches to Adaptation and Mitigation

6.4.1 Study and Draw up "Climate Change Law"

China has initiated a legislative process to establish a legal framework for defining responsibility and obligation of the whole society to address climate change (Zheng 2010). The legislation foresees two institutional mechanisms with one on international cooperation and the other on domestic coordination in addressing climate change.

Through the formulation of the "Climate Change Law," China will establish an open, transparent, and orderly information release system, define a unified publication system of scientific assessment reports and policy information on addressing climate change, and strengthen scientific research to combat climate change.

6.4.2 Strengthen Administration and Incentive Policies

Incorporate climate change mitigation and adaptation into China's national economy and social development plans. GHG emission control targets have already been broken down and assigned to provinces, and local implementation plans have been formulated; the climate change-related statistical system will be further improved, and the appraisal system for GHG emissions control targets are being gradually implemented.

Organize the formulation of special programs for addressing climate change. China has started to develop a national program specifically dedicated to climate change and stipulated priorities and major tasks for climate change mitigation and adaptation, low-carbon technology development, international cooperation, and other climate-related areas.

Fully carry out low-carbon pilot program. China launched a low-carbon pilot program in five provinces and eight cities. The pilot program is expected to accelerate the establishment of industrial systems, lifestyles, and consumption patterns characterized by low emissions. In the meantime, special pilot programs of low-carbon buildings, low-carbon transportation, and local effort to address climate change will be initiated along with the low-carbon pilot provinces and cities.

Actively explore the use of market mechanisms and economic instruments to control GHG emissions. China has continued to promote cooperation on the Clean Development Mechanism and published "Administrative measures for Voluntary Emission Reductions (VER) Transactions" regulating VER transaction activities to ensure an open, fair, and transparent market. In addition, China encourages and supports the pilot of carbon emissions trading along with the low-carbon pilot program and launched studies on the feasibility of imposing a carbon tax in China.

Formulate certification mechanisms for low-carbon products and undertake pilot programs for low-carbon certification. To control GHG emissions, China is formulating the "Administrative Measures for Low-carbon Products Certification," undertaking the pilot program of low-carbon products certification and encouraging the public to use low-carbon products, stimulating enterprises to upgrade the product structure.

Actively adopt policies and measures to adapt to climate change. China has taken effective measures in key sectors including agriculture, forestry and water resources, and in vulnerable areas, such as coastal zones, to seek benefits and avoid disadvantages, and minimize the negative impacts of climate change on economic and social development. At the same time, China will strengthen its efforts on monitoring, warning, forecasting of various types of extreme weather and climate events, release information in a timely manner, prevent and address extreme weather and climate disasters and their derivative hazards in a scientific way, and enhance the ability to cope with extreme weather events.

Adopt both mandatory and incentive mechanisms and policies. Based on a combination of various measures, China is setting a clear policy objective to promote actions to mitigate and adapt to climate change. Administrative policies include the total energy consumption control mechanism, emission control standards, access system, and appraisal system of the carbon intensity target; incentive policies and mechanisms include levying tax, carbon credits and carbon markets, and carbon trading and subsidies. Besides, voluntary mechanisms and policies, low-carbon product labeling and certification, and low-carbon consumption policies have been explored as well.

6.4.3 Increase Capital Investment

Establishing stable and sustainable fiscal policies and financing mechanisms, and supporting the research and development, application and commercialization of mitigation and adaptation technologies will be crucial for China fulfilling mitigation and adaptation targets.

In addition to increasing government financial investment, Central and local governments can institute other financial mechanisms to broaden financing channels and foster social, corporate, and private capital investment.

6.4.4 Strengthen Scientific and Technological Innovation

Increasing the overall administration on, policy guidance on, coordination of, and investment in scientific and technological innovation. It intensifies basic research on addressing climate change to enhance the ability of scientific judgment and accelerates the research and development, demonstration, and diffusion of major technologies for climate change mitigation and adaptation, especially energy conservation and efficiency improvement technologies, climate risk prediction, warning and adaption technologies, carbon capture and storage (CCS) technologies, and adaptation technology systems in sensitive areas. In addition, efforts should be made to introduce, digest, absorb, and reinnovate advanced technologies for climate change mitigation and adaptation.

6.4.5 Strengthen Capacity Building and Raise Awareness

Strengthen the personnel training in the field of climate change science and technology and foster a group of high-level and high-quality talents.

Disseminate energy conservation and emission reduction knowledge and raise the public's awareness of conservation. Strengthen publicity, educational, and training activities to disseminate climate change knowledge. Actively inspire the extensive participation by and conscientious action of social and civil organizations, and increase the public's interest and participation in decision-making and in management and supervision of actions addressing climate change.

6.4.6 Strengthen International Cooperation

China will continue to promote cooperation and exchanges with a pragmatic attitude, not only to learn experiences and introduce capital and technology from developed countries, but also to assist other developing countries, in the framework of "South–South Cooperation," to take actions to address climate change, especially least developed countries and most vulnerable countries (Xie 2010). China will also continue to strengthen multilateral and bilateral policy dialog and cooperation in the field of climate change.

References

Cai, F., Su, X., & Liu, J. (2008). The problems of coastal erosion and preventive measures under the background of global climate change. *Progress in Natural Science, 18*(10), 1093–1103.

Du, J., & Shi, Y. (1993). The effect of sea level rise on Jiangsu coastal hydraulic engineering. *Limnology, 24*(3), 258–279.

Hu, X., Liu, Q., & Jiang, K. (2007). Sectoral technological potentials to mitigate carbon emissions in China, (in Chinese with English abstract). *SINO-GLOBAL ENERGY, 12*(4), 1–8.

Hu, X., Zhang, A., & Liu, Q. (2011). *Abatement technologies and potential for energy end-use sectors in China* (Vol. 3, No. 24, pp. 382–407). Second Climate Change, National Assessment Report, Science Press (in Chinese)

Huang, Y. (2006). Emissions of greenhouse gases in china and its reduction staratery. *Quaternary Sciences, 5*(26), 722–732.

IEA. (2012). *World energy outlook 2012*. IEA publication.

IPCC. (2007). *Climate change: Mitigation*. Cambridge: Cambridge University Press.

Jiang, K., Xing, Z., Ren, M., & He, C. (2013). China's role in attaining the global 2 target. *Climate Policy, 13*(S01), S55–S69. doi:10.1080/14693062.2012.746070.

Ju, H., Chen, X-g, & Wang, T-m. (2011). Action framework for climate change adaptation: an agriculture case in Ningxia Hui nationality autonomous region. *Journal of Meteorology and Environment, 27*(1), 42–46. (in Chinese).

Liu, T., & Yan, T-c. (2011). Main meteorological disasters in China and their economic losses. *Journal of Natural Disasters, 20*(2), 90–95.

National plan on new-type urbanization (2014–2020) [EB/OL]. (2014-03-16) http://news.xinhuanet.com/house/wuxi/2014-03-17/c_119795674.htm.

Sun, C-y, Kang, X-w, & Ma, X. (2013). The situation and tasks of science and technology development on adaptation to climate change in China. *China Soft Science, 10*, 182–185.

Xie, Z. (2010). *Strengthen international cooperation to drive green development*, (Vol 10, pp 5–6) (in Chinese).

Zhang, X., & Wang, Z. et al. (2011). *Abatement technologies and potential for energy supply sectors in China* (Vol. 3, No. 24, pp. 366–381). Second Climate Change, National Assessment Report, Science Press (in Chinese).

Zheng, G. (2010). *Proposal for study and draw up "Climate change law"*. CPPCC Report, 26 Jul (in Chinese).

Chapter 7
Strategic Options to Address Climate Change

Jiahua Pan, Ying Chen, Haibin Zhang, Manzhu Bao and Kunming Zhang

Abstract China's rapid economic growth has led to greenhouse gas (GHG) emissions being ranked highest in the world. Although addressing climate change not only presents severe challenges, but it also presents opportunities, mitigation of, and adaptation to climate change that will allow China to meet its future development demands in a sustainable way.

Keywords Low-carbon development · UNFCCC · Urbanization

7.1 International Efforts to Address Climate Change, and Challenges and Opportunities for China

Over the past few decades, climate change has become not only a major international environmental issue, but also an important issue in international development, politics, and security.

J. Pan (✉) · Y. Chen (✉)
Institute for Urban and Environmental Studies, Chinese Academy of Social Sciences, Beijing 100028, China
e-mail: jiahuapan@163.com

Y. Chen
e-mail: cy_cass@163.com

H. Zhang
School of International Studies, Peking University, Beijing 100871, China
e-mail: zhanghb@pku.edu.cn

M. Bao
College of Horticulture and Forestry Sciences, Huazhong Agricultural University, Wuhan 430070, China
e-mail: baomanzhu@yahoo.com.cn

K. Zhang
Ministry for Environmental Protection, Beijing 100035, China
e-mail: kunminz1@gmail.com

D. Qin et al. (eds.), *Climate and Environmental Change in China: 1951–2012*, Springer Environmental Science and Engineering,
DOI 10.1007/978-3-662-48482-1_7

The need to understand and combat climate change has led to new scientific research and new technologies and also induced more competition between the major international powers. Climate change is also posing a threat to international security, due to sea-level rise, extreme weather events, water shortages, decreases in food production, and large-scale environmental immigration. Climate change may also present a challenge to the global governance model of the twenty-first century. Current international climate efforts are complicated by differences in the development phase of participating countries, and the political demands, geographical locations, ecosystems, and resources structures of each country. Overall, the participants in the effort to address climate change can be divided into two camps: the developing and developed countries. European Union, the USA, and China are among the key players in the international climate negotiations. The focus of these two camps is on historical responsibility for GHG emissions, financial assistance for mitigation and adaptation actions, and technology transfer.

The 17th Conference of the Parties (COP) of the United Nation Framework Convention on Climate Change (UNFCCC) held in Durban, South Africa, at the end of 2011 decided to establish the Ad Hoc Working Group on the Durban Platform for Enhanced Action (ADP) to develop a protocol, another legal instrument, or an agreed outcome with legal force under the Convention applicable to all Parties, no later than 2015, and for it to come into effect and be implemented from 2020.

China is facing growing international pressure to address its GHG emissions.

1. Industrial development in China is increasingly being constrained by global climate change efforts. Historically, the developed countries emitted large quantities of greenhouse gases (GHGs) during their industrialization process, without any constraints. However, since the 1990s, the situation changed significantly, and global climate governance has gradually strengthened. In 1990, under the framework of the United Nations, international climate change negotiations began seeking to allocate the limited global carbon space. This directly affected the pace of development in developing countries and had a major impact on.
2. China is finding its needs to balance its speed of economic development with maintaining its international image of a responsible power. China's rapid economic development has led to a significant upgrade in its international status, but also to a rapid increase in GHG emissions that is a concern to the global community. The international community has stated that China should take greater responsibility in the global response to climate change, and perhaps even play a leadership role. But China is still a developing country with a large population and a relatively low level of economic development and industrialization, dominated by coal in energy mix. In addition, China has complex climatic zones and fragile ecological environments.
3. With the growth of China's economic strength and international status, the international community has an increasing expectation for China to promote climate mitigation actions in developing countries in an effort to reduce their

emissions and deploy low-carbon technologies. It was stated in the UNFCCC that the Parties should protect the climate system for the benefit of present and future generations of humankind, on the basis of equity and in accordance with their common but differentiated responsibilities and respective capabilities. The developed parties should take the lead in combating climate change and the adverse effects by providing developing countries with financial assistance and technological transfer. Regrettably, there has been little progress since the Convention entered into force.

4. There has been an increasing difficulty in maintaining and strengthening solidarity, with respect to climate change, in developing countries. Within the camp of developing countries, there are inherent differences among different interest groups regarding the ability and priority in addressing climate change. In this context, China must try to maintain the unity of developing countries in order to maximize benefits in north–south cooperation.
5. China must also address the effects of low-carbon industries with respect to international trade. In recent years, some developed countries, under the pretext of protecting the climate, have advocated a border adjustment tax to be levied on imports of high-carbon products, the so-called carbon tariffs. The collection of carbon tariffs will sharply increase the cost of exports in developing countries and will have a serious impact on international trade.

Domestically, China is also facing many challenges in addressing climate change. First, the mitigation of climate change in China is facing economic constraints. Low-carbon technologies, as well as climate-friendly development policies and deployment and implementation of these technologies and policies, will increase economic costs and affect the existing policy framework. It was mentioned in the Fourth Assessment Report of The Intergovernmental Panel on Climate Change (IPCC) that there was a significant correlation between the total costs of emission reduction and the amount of emission reduction required in order to achieve a given target (IPCC WGIII 2007). Studies have shown that in China, the cost of marginal carbon abatement can be very high. When the reduction rate of carbon dioxide is between 0 and 45 %, the marginal carbon abatement cost is between 0 and $250 per ton of carbon dioxide. These costs will affect all levels of the economy, which will inevitably lead to some adjustment between the vested benefits and the expected benefits, and constrain GHG reduction activities.

The mitigation of climate change in China also faces domestic resource and environmental constraints. China's energy structure is currently dominated by coal. Coal accounts for 96 % of fossil fuel reserves, compared to only 4 % for oil and gas. In 2014, coal accounted for 66.0 % of total primary energy consumption and this directly contributed to China's high carbon dioxide emissions in recent years. According to the International Energy Agency (IEA 2013) estimates, China's carbon dioxide emissions are 35 % higher than that of the USA and will be 66 % higher by 2030. China's proportion of global emissions is projected to increase from 19 % in 2005 to 27 % in 2030. It is difficult to change China's coal-dominated energy structure in the short term, and the process of transition to a low-carbon development model will force China to take on more financial and technical burdens than many other countries.

Some of China's current laws, strategic planning, and policies also affect China's ability to mitigate climate change. At present, the development of a low-carbon economy is in its infancy, and there is a need to improve implementation and integrate climate policy into development strategy and macroeconomic policies. In addition, there are still many institutional barriers to transit toward a low-carbon economy, such as a lack of incentives for low-carbon products and technologies, and excessive subsidies of high-carbon energy production.

However, China also has rare opportunities to respond to climate change. A successful response to climate change may help adjust China's economic structure, get rid of the carbon lock-in effect, and allow China to realize the economic gains. One key element to unlock high-carbon development pathways lies in the large-scale development and use of alternative energy technologies (based on low-carbon or zero-carbon energy sources) in everything from energy extraction, production, and transportation, to final consumer systems. In a sense, industrialization in developing countries today is largely repeating Western developed countries' carbon lock course in history, causing a global carbon lock: The heavy carbon industries and technologies of the developed countries are transferred to the industries of developing countries through international investment and trade channels. However, the new low-carbon world economy will eventually enable developing countries lessen their carbon lock. In such a competitive system, China will have an opportunity to invest in more innovation, rather than simply imitating the technology of the developed Western countries, or blindly accepting industrial technology transfer from developed countries; these efforts can change China's historical path of "treatment after pollution." China may even be able to leap forward in technology, to achieve the transition to a low-carbon economy better and faster than in developed countries.

Addressing climate change may help China to strengthen the competitiveness of its low-carbon products in the international market and to occupy a larger share of international markets. China has the potential to become the world's largest green energy market, the largest producer of low-carbon goods, and the largest exporter of low-carbon products. China currently has a number of low-carbon products, such as energy-saving lamps, photovoltaic power generation equipment, and wind power equipment, with a huge export potential.

While China increases its own innovation and its ability to create low-carbon technologies, as a developing country China should seize the opportunity to urge developed countries to transfer low-carbon technologies, thereby strengthening international cooperation. This will help to speed up the global pace of development of a low-carbon economy. China has also made efforts to promote south–south cooperation with other developing countries, especially in Africa and in Latin America in variety of fields including low-carbon technology transfer and deployment.

7.2 Domestic Reasons for China to Address Climate Change

China needs to address climate change, not only to adapt to the global marketplace, but also to address domestic political, economic, social, and environmental sustainable development.

7.2.1 *Adverse Impacts of Climate Change on China's Fragile Ecological Environments*

Climate change is already having adverse impacts on China's fragile ecosystems, such as farming-pastoral regions, northern forests, alpine regions, and river headwater and drainage regions.

The impacts of climate change on forest ecosystems mainly show up in the forest ecosystem structure, composition, and distribution of vegetation. Climate change has an influence on forest productivity and carbon cycle functions, and therefore affects the biogeochemical cycle. Climate change can decrease the biodiversity of ecosystems, resulting in the loss of many valuable forest tree species. In addition, the climate-related increase in intensity and frequency of extreme weather events lead to a decrease in forest ecosystems. The prairie/grassland ecosystem is one of the principal ecosystems in China and mainly occurs in arid and semiarid areas of the country. Climate change may lead to increased aridity in the northern regions of China, resulting in natural grassland degradation and desertification and decreased grassland production. The grasslands of Inner Mongolia and Qinghai-Tibet Plateau are some of the most vulnerable areas in China.

Climate change affects the hydrological cycle of wetlands. In recent decades, the wetlands of northeastern China have been threatened by climate warming. Sanjiang Plain has experienced a dramatic reduction of wetland areas due to exceptional climate change in the region. Wetland changes are negatively correlated with temperature change, and positively correlated with rainfall and humidity changes. These changes have also been seen in the source area of the Three Rivers. Between 1990 and 2004, the climate at the source area of the Yellow River changed drastically, faster than global climate change, which resulted in a shrinking of the source area.

Desert ecosystems are divided into stone or gravel goby deserts and sandy deserts. These areas have poor biological diversity, low population density, and low richness of vegetation, so they are very fragile. China has large desert areas and has experienced an increase in desertification across other regions since the 1950s. For example, the MU US sandy land region of Inner Mongolia has been expanding since the 1950s, due to reduced precipitation (Na et al. 1997). Future climate warming and drying will further increase desertification in this region. Desertification threatens ecological security and the sustainable economic and social development of the affected regions.

With recent warming of the Tibetan Plateau, permafrost has degraded in seasonally frozen areas, while evaporation in desert areas of the Plateau has increased, making the area more arid and leading to the aggravation of desertification. In 2005, the total area of rocky desert in China was 129,600 km^2 (Karst Rocky Desertification Situation Bulletin 2007), mainly located in Guizhou, Yunnan, and Guangxi provinces or autonomous regions, and accounting for 67 % of the rocky desert of the whole nation. China's rocky deserts are growing at a rate of 3–6 % per year. Climate change in karst regions can lead to an increase in rainstorms in spring and summer, and increased soil erosion, thereby accelerating the rate of rocky desertification.

China has the highest number of mountainous region disasters in the world, such as landslides and debris flow. Due to climate change, the number of days of rainfall in China has decreased overall, but the intensity of rainfall events has increased, resulting in a type of stabilization. However, intense rainfall events can suddenly increase the moisture content of the soil, leading to an increase in the likelihood of debris flows and landslides.

7.2.2 Reasons to Promote Industrialization and Urbanization for Low-Carbon Transition

The major Western developed countries are currently in the post-industrialization stage, with the construction of their large-scale infrastructure being generally complete. Their national economies mainly rely on high-technology and services industries, and energy consumption has therefore stabilized or even dropped. China is currently in the middle stages of industrialization. Since 2006, the proportion of heavy industry has accounted for ~70 % of the total industrial economy. As the development of energy-intensive industries, such as steel milling, automobile manufacturing, and shipbuilding, accelerated, GHG emissions rapidly increased in China. However, the level of industrialization has varied across the country. The eastern region of China is currently in the later stages of industrialization; the northeast regions are in the middle of industrialization; and the central and western regions are in the latter half of the pre-industrial stage (Chen et al. 2006). So, while China has experienced some great economic success, there is still extreme poverty in the countryside. In the process of industrialization, rapid economic growth is needed to meet development demands and reduce poverty; however, unfortunately, GHG emissions will continue to increase in the future.

Urbanization occurred accompanying with industrialization. At the end of 2011, China's urbanization rate reached 51.2 %. It is predicted that by 2030, the urbanization rate may reach more than 65 %, and the urban population will reach approximately one billion (Ni 2009). Accelerated urbanization will require large-scale expansion of urban infrastructure. Roads, electricity, water supply, gas supply, public transport, municipal facilities, cultural and recreational facilities, sanitation facilities, and other infrastructure construction will generate a huge demand for energy-intensive products such as steel, cement, and chemicals. Urbanization also leads to changes in patterns and structure of energy consumption. One important task of China's modernization process is to narrow the gap between urban and rural areas, with respect to infrastructure, economic success, and energy consumption. Improving rural infrastructure is important to raise rural living standards, but this is bound to create an increase in GHG emissions from energy-intensive industries.

These trends of industrialization and urbanization are driving the efforts to pursue low-carbon development in China.

7.2.3 Socioeconomic Reasons to Pursue Low-Carbon Development

Mitigation actions not only help to reduce the direct impacts of climate change on China's long-term social and economic development, but also being consistent to China's efforts to protect the ecological environment and ensures energy security. The improvement of energy efficiency and the development of renewable energies will ease the pressures on China's energy supply, and a reduction of the use of coal and other fossil energy will reduce emissions of sulfur dioxide (SO_2) and nitrogen oxides (NO_x) and reduce air pollution (such as acid rain and suspended particulates).

Addressing climate change can also contribute to technological innovation in China. The urgent need to address climate change can promote the development of new low-carbon technologies and new renewable energies. With the help of international technology transfer efforts, and by strengthening its own research and development, China can increase its technology capabilities and enhance its international competitiveness (Wang and Zou 2009).

Addressing climate change also includes the transformation of Chinese consumption patterns. The urban residents' contribution to carbon dioxide and other GHG emissions is much higher than that of rural area. It is estimated that between 1999 and 2002, urban energy use accounted for around 26 % of the annual total energy consumption and 30 % of the carbon dioxide emissions in China. In the USA, urban energy consumption in transportation and residential use accounts for approximately 66 % of the total energy consumption. Therefore, changing the sustainable consumption patterns is also very important to address climate change.

Addressing climate change in China will bring significant business opportunities, contribute to economic restructuring, and enhance competitiveness. Low-carbon technologies development, low-carbon goods production and services will create new markets, bringing new driving forces for economic growth, and creating new employment opportunities. According to Roland Berger (Roland-Berger Strategy Consultants), the global environmental goods and services market is expected to double by 2020, on the basis of $1.37 trillion per year, reaching $ 2.74 trillion; half of this will be due to the energy efficiency market. In Europe and the USA, increased investment in the energy efficiency of buildings will add 2.0–3.5 million green jobs. In developing countries, the potential for green jobs may be larger.

HSBC studies show that the profits of companies in the global climate change-related industry in 2008 (including renewable energy power generation, nuclear energy, energy management, water treatment, and waste disposal enterprises) reached $534 billion, exceeding the total profits of the aerospace and defense industries (of $530 billion), and low-carbon industries are becoming the new pillars of the global economy (Ju 2010). China is poised to exploit its advantages in creating low-carbon technologies.

7.3 China's Strategic Response to Climate Change

China attaches great importance and developed a series of strategies and policies to address climate change. It was emphasized in Hu Jintao's report of the 18th CPC national congress that we must give high priority to making ecological progress and incorporate it into all aspects and the whole process of advancing economic, political, cultural, and social progress, work hard to build a beautiful country, and achieve lasting and sustainable development of the Chinese nation. In 2005, the National Economic and Social Development of the Eleventh Five-Year Plan clearly proposed that energy consumption per unit of GDP should be decreased by 20 % in 2010 below the level in 2005, and emissions of major pollutants should be reduced by 10 %. China is striving to build a resource-conserving and environment-friendly society. In 2007, the Chinese government issued China's National Climate Change Program. It was the first comprehensive policy document to address climate change, with specific goals, basic principles, and focus areas and policy measures to address climate change by 2010. In the same year, the Chinese government set up a National Leading Group on Climate Change, to guide departments and local governments to promote energy conservation. Promulgated in 2010, the 12th Five-Year Plan clearly emphasized the need to actively respond to climate change, and in this plan, for the first time, a target was proposed to reduce carbon emissions per unit of GDP by 16 % in 2015 below the level in 2010.

China needs to find a low-carbon development path in line with its specific circumstances. To achieve China's low-carbon development goals demands, great efforts in adjustment of the current energy structure, development of renewable energy, and reduction of fossil fuels consumption, are needed. The strategic options for transition to a low-carbon society should include the following aspects:

First, mainstreaming climate policy into the general framework of sustainable development: China needs to continue to vigorously promote energy savings, raise awareness of the whole society on low-carbon development, and make it a common goal of government agencies, social organizations, enterprises and institutions, and individuals.

Second, building a low-carbon economic system: The National economic system should accelerate the construction of low-carbon systems in industry, agriculture, transportation, construction, and service industries.

Third, building low-carbon consumption patterns: China should advocate sustainable and low-carbon consumption system, promoting low-carbon lifestyles, resource-saving and low-consumption patterns, environmental protection, and the production of green products.

Fourth, building governance system to promote green and low-carbon development: China's government must see green and low-carbon development as an important goal; it must promote green development as one of the most important functions of government and a vital indicator for assessing the work of governments at all levels. The Government needs to set up specialized agencies to fully promote and carry out green development programs. In addition, the wide participation of the public also plays an important role in promoting low-carbon development.

In short, China must mobilize its international and domestic capabilities actively in the international arena and in the domestic level not only to promote green and low-carbon development in China but also to make its contribution to global efforts to address climate change.

References

Chen, J., Huang, Q., & Zhong, H. (2006). The synthetic evaluation and analysis on regional industrialization. *Economic Research Journal, 6*, 4–15.

China Central Government. (2007). *Karst rocky desertification situation bulletin* (Vol. 8, pp. 58–60) July 15, 2007. Available at http://www.gov.cn/ztzl/fszs/content_650610.htm

International Energy Agency. (2013). *CO_2 emissions from fuel combustion highlights 2013*. Available at http://www.iea.org/publications/freepublications/publication/name,43840,en.html

IPCC. (2007). *Climate change 2007: Mitigation: Contribution of working group III to the fourth assessment report of the intergovernmental panel on climate change*. Cambridge and New York: Cambridge University Press.

Ju, J. (2010). *Some considerations and perspectives of world and national economy* (No. 7). USA: People's Tribune.

Na, P.-s., Wang, Y.-k., Man, D.-l., & Xu, S.-l. (1997). Study on the eco-environment of MU US sandy land. *Journal of Desert Research, 17*(4), 410–414. (in Chinese with English abstract).

Ni, P. (2009). Beijing: Urban environment and competitiveness. *China Business and Market, 8*, 015. (in Chinese with English abstract).

Wang, H., & Zou, J. (2009). Study on technology transfer and emissions control and reduction in developing countries. *Journal of Environmental Protection, 4*(12), 74–77.

Chapter 8
Concluding Remarks

Dahe Qin, Yong Luo, Guangyu Shi, Yongjian Ding, Wenjie Dong, Erda Lin and Jiahua Pan

Abstract This chapter brings together key information contained in "Evolving Climate and Environment in China: 2012," a Chinese monograph. Significant warming in China for the last 100 years is unequivocal, confirmed by substantial observational data. Anthropogenic activities are very likely the key driver of the

The following researchers have also contributed to this chapter

Xiulian HU

Energy Research Institute, National Development and Reform Commission, Beijing, China
e-mail: huxl@eri.org.cn

Xiaoye Zhang

Chinese Academy of Meteorological Sciences, China Meteorological Administration, Beijing, China, e-mail: xiaoye@cams.cma.gov.cn

Kejun Jiang

Energy Research Institute, National Development and Reform Commission, Beijing, China, e-mail: kjiang@eri.org.cn

Shaowu Wang

Department of Atmospheric and Oceanic Sciences, School of Physics, Peking, University, Beijing, China, e-mail: swwang@pku.edu.cn

Xuejie Gao

Institute of Atmospheric Physics, Chinese Academy of Sciences, Beijing, China
e-mail: gaoxj@cma.gov.cn

D. Qin (✉)
China Meteorological Administration, Beijing 100081, China
e-mail: qdh@cma.gov.cn

Y. Luo (✉)
Center for Earth System Science, Tsinghua University, Beijing 100084, China
e-mail: yongluo@tsinghua.edu.cn

G. Shi
Institute of Atmospheric Physics, Chinese Academy of Sciences, Beijing 100029, China
e-mail: shigy@mail.iap.ac.cn

D. Qin et al. (eds.), *Climate and Environmental Change in China: 1951–2012*,
Springer Environmental Science and Engineering,
DOI 10.1007/978-3-662-48482-1_8

warming in China since the 1950s. Climate change has significantly impacted the eco- environment and socio-economy of China. The fast increase in greenhouse gas emissions from China is due to the rapid growth of China's economy and its position in the world economy. Climate change projects show warming in China will continue throughout the twenty-first century. Technological and policy options for adaptation and mitigation are critical to China's actions on climate change. Green industrialization and low-carbon urbanization are essential for sustainable development. The future research priorities and goals are also summarized.

Keywords Concluding remarks · Climate change · China · Impact · Adaptation · Mitigation · Option · Research priorities

This chapter brings together key information contained in "Evolving Climate and Environment in China: 2012," a Chinese monograph consisting of four volumes: scientific basis, impact and vulnerability, climate change mitigation, and future research priorities and goals. This information will be a useful resource to decision makers, professionals, and others who are concerned about climate change issues.

8.1 Conclusions

8.1.1 Significant Warming in China for the Last 100 Years Has Been Confirmed by Substantial Observational Data

Warming in China over the past century has been clearly evidenced by observational data, such as rising average land surface air temperatures and offshore sea

D. Qin · Y. Ding
State Key Laboratory of Cryospheric Sciences, Cold and Arid Regions Environmental and Engineering Research Institute, Chinese Academy of Sciences, Lanzhou 730000, China
e-mail: dyj@lzb.ac.cn

W. Dong
Zhuhai Joint innovative Center of Climate-Environment-Ecosystem, Beijing Normal University, Beijing, China
e-mail: dongwj@bnu.edu.cn

E. Lin
Institute of Environment and Sustainable Development in Agriculture, Chinese Academy of Agricultural Sciences, Beijing 100081, China
e-mail: lined@ami.ac.cn

J. Pan
Research Centre for Sustainable Development, Chinese Academy of Social Sciences, Beijing 100028, China
e-mail: jiahuapan@163.com

surface temperatures, widespread shrinkage of the cryosphere, and the gradual rise of mean sea level along China's coasts.

Meteorological measurements since A.D. 1880 show that the average land surface air temperature has increased significantly in China, at a rate of 0.5–0.8 °C per 100 year, in particular since 1980s. The highest average land surface air temperature ever recorded occurred in 2001–2010. From 1980 to 2011, air temperatures and sea surface temperatures both increased in coastal China, by 0.4 and 0.2 °C per 10 year, respectively, with the most significant warming found in continental shelf seas.

The average minimum air temperature in China has increased by 0.5–0.6 °C per 10 year. Low-temperature-related extreme events significantly weakened and/or declined over this period. From 1961 to 2010, the number of cold waves decreased significantly, by 0.3 per 10 year, while frost days decreased by 3 days per 10 year. High-temperature events (such as heat waves) showed remarkable decadal fluctuations.

Precipitation in China shows an inter-decadal variability of 20–30 year. Since the 1980s, winter and summer monsoons have been weakening in eastern Asia, which consequently alters the occurrence of floods and droughts in China. There have been a significantly increasing number of regions in China with high numbers of intense precipitation events, while the amount of overall precipitation and the frequency of wet spell are decreasing. Drought patterns show the marked regional differences across China, with more arid climate and droughts found in the north, northeast, and eastern northwest areas of the country.

Since 1951, extreme weather phenomena, such as typhoons and extratropical cyclones, sand and dust storms, high winds, and thick fog, have tended to decrease or weaken in China. The average intensity of tropical cyclones in China has not changed significantly. The intensity of strong tropical cyclones has decreased, while typhoons with landfalls in China have proportionally increased. Cold air masses have weakened, leading to a reduced average wind speed, and a smaller number of windy days, and hence a significantly reduced occurrence of dust storms in northern China. In most parts of China, foggy days have been decreasing, while regional hazy weather has been increasing.

Modern glaciers in China have retreated and thinned since the 1960s; permafrost temperatures have increased, the active layer has thickened; seasonal frozen ground has thinned and shrunk in area; snow-covered areas have slightly decreased; and the number of days with snow cover has significantly decreased.

The mean sea level of coastal China has significantly risen. From 1980 to 2011, it rose by 2.7 mm/year, a rate that is higher than the global average for the same time period. From a global perspective, acidification in China's coastal waters is moderate; the pH value in most seawater in the 1990s was 0.06 less than that in pre-industrial times.

8.1.2 Human Activities Are Very Likely the Key Driver of Warming in China Since 1950s

New observations and climate modeling show that climate change in China since the middle of the last century is very likely to be caused by intensified human activities worldwide, such as increased greenhouse gas (GHG) emissions and changes in land use and land cover. Natural factors alone fail to explain recent warming across China. For example, the radiative forcing caused by human activities since A.D. 1750 is 1.6 W/m^2, while the driving force (radiative forcing of +0.12 W/m^2) behind the climate change generated by solar variability in the past 100 year is less than one-tenth of that of human activities. Volcanic eruptions generally result in lower surface air temperature, and these effects last for no more than several years. Geothermal heat flow contributes only 1/18 the warming effect of anthropogenic forcing factors. Natural climate cycles and air–ocean interactions on inter-annual and decadal time scales affect climate, but this effect is difficult to determine in terms of its contribution to the changing global climate on a time scale of more than 100 year.

8.1.3 Climate Change Has Had Significant Impacts on China's Natural Environment and Its Society and Economy

Climate change has had significant impacts on the water resources, ecology, agriculture, health, society, and economy of China.

Since 1950, the measured runoff of China's major rivers has generally declined, and the quality of water resources has also declined due to increased water pollution. Shrinking glaciers have resulted in an increase in mountain runoffs in arid regions, by 5–30 %. The climate-related changes in frozen ground have altered the mountain runoff cycle, with melting snow running off about one month earlier. Rising sea level has further eroded the coasts of China and caused seawater intrusion into freshwater sources.

Climate change has affected the structure, composition, distribution, phenology, and productivity of forests. Grasslands in China are degrading by 2×10^6 ha every year, caused in part by the aridification of northern China. Significant changes have also been observed in wetland hydrology and ecology, lake waters, and plant communities.

Climate change has affected farming, animal husbandry, and fisheries in varying degrees. The increased air temperatures have led to a northward shift of agro-climatic zones, the prolongation of growth periods, the extension of multi-crop boundaries northward and to higher altitudes, the aggravation of heat waves and droughts, and the intensification of biohazards. Temperature impacts on livestock products are positive or negative, but the negative effects outweigh the positive effects. Climate change has also led to declining fisheries in China.

Human health is vulnerable to high-temperature and extreme weather events. For example, the distribution of malaria and schistosomiasis has changed across China. The heat island effect is affecting urban citizens both directly and indirectly. Climate change requires higher construction standards for transportation infrastructure (e.g., roads and railways) and for urban infrastructure. Construction projects affected by climate change in China include the Three Gorges project, the South-to-North Water Diversion project, the Yangtze Estuary Renovation, the Qinghai–Tibet Railway, the West-to-East Gas Transmission project, the China–Russia Oil Pipeline, and the Three-North Shelterbelt project. The warming climate in China coincides with a significant increase in cooling-related energy consumption as well. The impact of climate change is being increasingly felt by regional tourism (changing landscapes and tourist seasons).

It is difficult to completely isolate the impact of climate change on economic and social systems from natural changes, due to the intertwining of climate adaptation action and non-climatic drivers. At a regional or smaller scale, it is difficult to quantitatively attribute the modeled and observed temperature changes to natural or human causes.

8.1.4 Rapid Growth of the China's Economy and Its Position in the World Economy Have Led to a Fast Increase in Greenhouse Gas Emissions

Since 2000, GHG emissions have increased significantly in China, along with the booming economy and the country's emerging position as a "World Factory" in the international economic landscape. According to *The Long-term Trend in Global CO_2 Emissions—2011*, a report released by the Netherlands Environmental Assessment Agency and the European Commission's Joint Research Center, CO_2 emitted from global energy and cement production processes was 33 billion tons in 2010, of which 9 billion was emitted by China, 5.2 billion by the USA, 4 billion by the European Union, 1.2 billion by Japan, and 1.8 billion by India. The CO_2 emissions from energy and cement production processes in China have doubled since 2003 (Olivier et al. 2011).

According to the scenario studies from several groups, China will demand significantly more energy for its further economic development, which will result in a continued increase in its CO_2 emissions. Scenario studies show that by 2030, China will need 50–100 % more energy than in 2008, and by 2050, 85–150 % more. By 2030, China will emit over 11 billion tons of energy-related CO_2, and 12–17 billion tons by 2050.

Models also show, however, that China has a great energy-saving potential. If current energy conservation policies and measures are strengthened, after 2020

China could have a flat energy demand. By 2030, it might be possible to control energy demands and use less than 4.8 billion tons of standard coal (less than 5.5 billion tons by 2050). If climate change adaptation policies are instituted, the country's CO_2 emissions may fall to 8.5–11 billion tons by 2030, and to 5–8 billion tons by 2050. To achieve this, China needs to accelerate the development of renewable energy and nuclear power and accelerate the application of energy efficiency and carbon capture and sequestration technologies, and expand energy-saving, emission-reduction, and low-carbon policies.

8.1.5 Climate Warming in China Will Continue Before the End of the Century

Over the past five years, international climate model groups have developed climate system models with the higher resolutions and better expressions of geophysical, chemical, and biological processes; these improve the credibility of future climate change projections. The international scientific community has adopted the new representative concentration pathway (RCP)-based GHG emission scenario. Five models developed in China are being used in the Coupled Model Intercomparison Project Phase 5 (CMIP5), lending support to the International Panel on Climate Change (IPCC) Fifth Assessment Report (AR5) (IPCC 2013).

The latest CMIP5 multi-model estimates that under scenario RCP 2.6, by the end of the twenty-first century, the global average surface temperature will increase by 0.6 °C (−0.1 to 1.3 °C) above that of 1986–2005, and global precipitation will increase by 0.06 mm/day (0.04–0.08 mm/day), in turn, the temperature and precipitation in China will increase by 1.4 °C and 5.6 %, respectively.

Under scenario RCP 4.5, the global average surface temperature will probably increase by 1.6 °C (0.8 to 2.5 °C) and global precipitation will increase by 0.1 mm/day (0.07–0.14 mm/day), while the temperature and precipitation in China will increase by 2.5 °C and 8.8 %, respectively.

Under scenario RCP 8.5, the global average surface temperature will increase by 4.4 °C (2.7–5.6 °C) and global precipitation will increase by 0.19 mm/day (0.14–0.23 mm/day), while the temperature and precipitation in China will increase by 5.1 °C and 13.5 %, respectively. Compared with the projected temperature changes, the projected precipitation has a much higher uncertainty. The temperature increase in China is projected to be higher than the global temperature increase.

Climate change will lead to the alteration of agriculture in China. The growing belts of crops such as wheat and corn will move northward, and wheat and rice production will drop. Taking into account CO_2 effects, wheat and rain-fed corn production will rise, while irrigated rice production will drop slightly. In the future, boreal forests, temperate deciduous and evergreen forests, and tropical forests will move noticeably northward. Tropical dry forests and savannas will expand, and the tundra on the Qinghai–Tibet Plateau will shrink. Climate change will reshape the vegetation coverage in China by 39–49 %, mainly in the transition zone extending from northeastern to southwestern China, and the transition zone extending from the woodlands in the east to the grasslands in the west.

River runoff will increase in China, except for areas with reduced runoff in Ningxia, Jilin, Hainan and Shanxi.. Overall, however, due to factors such as increased evaporation caused by warming, water availability issues will increase, such as the water shortage plaguing the Haihe River and Yellow River basins, and the water over-availability found in Fujian, Zhejiang, the middle and lower reaches of the Yangtze River, and the Pearl River basin. Sea level rise caused by climate change and by human activities such as groundwater overexploitation in particular, will submerge coastal land, intensify storm surges, and worsen coastal erosion and estuarine seawater intrusion.

The multi-CMIP5 global model performs well in simulating air temperature in China. Due to its low resolution, however, the model is less reliable when simulating temperatures over areas with high topographic variation. In addition, the simulated temperature in the Qinghai–Tibet Plateau is deemed rather low. In terms of the precipitation, the model simulates a basic distribution of high precipitation in the south and low precipitation in the north and northwest. At the same time, however, a false precipitation center is found in the eastern Qinghai–Tibet Plateau. The uncertainty in future climate change projections arises from differences across climate models in describing the processes of cloud feedback, ocean heat uptake, and carbon cycle feedback, as well as from the uncertain estimates of the impacts of GHG-related policies, future population growth, economic growth, technological progress, and new energy development and management restructuring.

8.1.6 Technological and Policy Options for Adaptation and Mitigation Are Critical to China's Actions on Climate Change

1. Technological options for adaptation

China should continue to modify its agricultural structure and cropping systems in response to the changing climate, in order to deal with extreme weather and climate events and their secondary and derivative disasters. China should place an emphasis on the multi-level reuse of material energy and the recycling and regeneration of waste.

It is necessary to incorporate climate change into the water assessment and planning process in China so that water will be managed in an integrated manner, with improved allocation, better water management information systems, and improved water projects.

It is necessary to further enhance reforestation and forest management to better prevent and control forest fires, pests and plant diseases; to moderate the existing intense grazing of pastures and enlarge irrigated grasslands and artificial pastures; to protect wetlands and control river and lake pollution; to conserve desert biological resources, combat desertification, and strengthen the conservation of species; and to increase efforts in the breeding of rare and endangered species; and to enhance the adaptability of ecosystems to climate change.

China will need to improve the design of tidal control infrastructure in coastal areas, including increasing the investment in the construction of seawalls and river banks, and reinforcing existing coastal shelterbelts. Monitoring of ground subsidence needs to be improved, and the standards on recharged wastewater strengthened, in addition to groundwater level observations and groundwater quality monitoring. China must regulate groundwater exploitation, focusing on shallow groundwater, and increase the utilization of salt water where appropriate.

Climate change will affect the distribution and frequency of tropical diseases and affect human health with regard to heat and cold stress, etc., and so, China must develop a sound monitoring and warning system to detect climate change impacts on human health.

2. Technological options for mitigation

New coal-fired power plants in China need to be further upgraded with the integrated gasification combined cycle (IGCC) technology. China also needs to increase the use of hydropower as appropriate, develop safe nuclear power, scale up wind power, and boost solar photovoltaic and biomass power generation. Systems and technologies such as IGCC, poly-generation, and CO_2 capture and sequestration may be promising options in the medium and long term.

Technologies such as by-product and waste recycling, raw material and fuel substitution, and CO_2 capture and storage (CCS) are all important options for CO_2 reduction, especially when used in cement and steel production. Increased fuel economy standards, improved efficiency of internal combustion engines, hybrid power vehicles, and plug-in hybrid electric and fuel-cell vehicles all contribute to a substantial emission reduction in the field of transportation. New technologies for heating, efficient lighting, energy-efficient appliances, insulation, and solar thermal utilization are essential in achieving cost-effective CO_2 reductions. In the long run, however, it will be necessary to deploy low-carbon and zero-carbon technologies such as heat pumps, solar thermal systems, cogeneration systems, and on-the-spot renewable-energy-based power generation in order to reduce CO_2 emissions in China.

Recently, China has witnessed a dramatic drop in the cost of selected emission-reduction technologies, especially advanced coal-fired power generation, onshore and offshore wind power generation, and photovoltaic power generation. These technologies, together with originally low-cost hydropower and nuclear power technologies, will help to lower the cost of CO_2 emission reduction. Meanwhile, CCS technology in China is significantly more affordable than other technologies and is being deployed in the project demonstration phases.

3. Policy options

China must develop an overarching policy framework on for no-carbon and low-carbon technology development. There needs to be a single clearing house for scientific assessment reports and policy information on climate change and transparent publication of these reports and policies. China also needs to strengthen its scientific and technological research on climate change.

Climate change mitigation and adaptation must be important elements in China's national programs for economic and social development. China needs to identify targets and priorities in national GHG emission reduction, develop analysis and evaluation mechanisms, and pilot low-carbon initiatives.

China should use economic policies and market mechanisms to introduce mandatory and incentive mechanisms for GHG emission control. The country should implement carbon pricing measures as soon as possible, including carbon taxes and carbon trading. A low-carbon certification system should be piloted, and policies implemented that allow for better financing and subsidies of low-carbon systems in order to motivate the general public in low-carbon consumption.

China must take proactive measures and policies to adapt to climate change. It is necessary to strengthen the monitoring, warning, and forecasting of extreme weather and climate events in such key sectors as agriculture, forestry, and water resources so that information can be released in a timely fashion and extreme weather events can be dealt with more effectively.

China needs to improve technological innovation, and research and development. The country should further strengthen scientific research, technological development, and scientific and technological service delivery on climate change, mobilizing the enthusiasm and creativity of scientists and technologists to engage in such research and commercialization.

International cooperation is critical in addressing climate change. The Chinese government will continue to actively and pragmatically participate in and promote technology transfer and international cooperation, under the United Nations Framework Convention on Climate Change, and actively engage in cooperation with developed countries on clean development mechanism (CDM) projects.

8.1.7 Green Industrialization and Low-Carbon Urbanization Are Essential for Sustainable Development

Addressing climate change presents both challenges and opportunities. First, China must have a well-defined strategic goal for addressing climate change that covers targets for mitigation and adaptation from both energy security and climate security perspectives. Second, climate change mitigation and adaptation, which cannot be isolated, must be incorporated into the overall framework of sustainable development, taking into consideration the level of economic development, constraints of resources and environment, and the needs of society. Third, it is necessary to strengthen technological innovation; accelerate research and development, and commercialization of advanced low-carbon technologies; and increase investment in low-carbon development by stepping up emerging strategic low-carbon businesses, and transforming and upgrading traditional businesses. Fourth, as part of the world economy, China needs to be mainstreamed into the overall process of international cooperation in climate change. Fifth, it is necessary to encourage the concept of low carbon consumption in society. Sixth, it is necessary to develop a sound legal, regulatory, and institutional framework to create policies and markets for low-carbon development. As its national characteristics determine, China must lead the low-carbon transformation efforts for vulnerability reduction and climate change adaptation and promote the global process on climate change.

8.2 Future Research Priorities and Goals

8.2.1 Science Basis

In order to further reduce key uncertainties in the scientific understanding of climate change, it is necessary to make in-depth studies of the processes and mechanisms of climate change, and of its causes, impacts, and future trends.

1. Integration of multi-source data on the climate system

It is necessary to study how to collect, assimilate, and integrate climate observations, including integration of multi-source and multi-scale data, as well as collection and analysis of related socio-economic data.

2. Reconstruction of past climate change data series with high-precision

It is necessary to develop new theories, methods, and technologies for the reconstruction of past climate change, to study how to integrate various proxy data of climate change records, and to compare and analyze past warm periods.

3. Study of the behavior and mechanisms of global climate change

It is necessary to analyze the behavior of climate change and its characteristics, including natural climate variability and the processes and mechanisms of anthropogenic impacts on climate change.

4. Development of earth system models, simulation, and projection of climate change

It is necessary to develop and improve climate system models, by studying key parameterized physical, chemical, and biological processes and their uncertainties, as well as techniques for coupling different component models. This will allow the development of the improved numerical models and projects of climate change.

8.2.2 Impact and Adaptation

Key environmental and economic areas of water resources, agriculture, forestry, ocean systems, human health, ecosystems, and disaster management must be emphasized in China's efforts to improve research on the mechanisms and impacts of climate change. This includes enhancing research and development of adaptation theory and technologies, initiating adaptation demonstration projects in typically vulnerable regions and fields, and actively factoring the climate change response into regional sustainable development plans.

1. Mechanism of climate change impacts and assessment

China must improve research into the impact mechanisms of climate change and extreme climate events, and into integrated assessment models, to analyze the current and future risks of climate change and global warming.

2. Adaptation theory and technological research and development

Climate adaptation technologies should be studied by all sectors and regions in China. This will aid in the development of adaptation decision-making processes; the assessment of capital and technology requirements, the development of specific adaptation technologies for vulnerable areas; the development of technologies addressing extreme climate event preparedness and disaster management, and the integration, application, and extension of adaptation technologies across social and economic sectors.

3. Typical vulnerable sectors and regional adaptation demonstrations

China has vulnerabilities to climate change in the areas of agriculture, forestry, water resources, human health, biodiversity and ecosystems, economic and development projects, and disaster management. These areas need to be highlighted in the study of adaptation strategies and measures. Moreover, it is necessary to make a cost–benefit analysis of such actions and measures and demonstrate climate change adaptation technologies.

4. Climate change adaptation and regional sustainable development

It is necessary to identify regions and groups vulnerable to climate change, and the priorities for their adaptation to such change. This will aid in the integration of climate change adaptation into regional economic and social development programs, including those in megacities and underdeveloped areas, and will assist policy makers with climate change adaptation measures and plans for international cooperation in climate change adaptation.

8.2.3 Mitigation

The scientific and technological support of the mitigation of GHG emissions, and the promotion of a low-carbon economy needs to be emphasized in order to facilitate the innovation and marketing of non-fossil energy sources and clean coal technology; improve energy efficiency and the use of advanced low-carbon technologies in the industry, construction, and transportation activities; drive the research and development of key technologies in such fields as forestry carbon sinks and industrial carbon sequestration; address the cost reduction and commercialization of such key technologies as carbon capture, utilization, and storage; and propose priorities and goals in future climate mitigation research.

1. Clean energy and clean coal technology

China needs to develop renewable-energy power generation technologies such as cost-effective offshore wind power, solar photovoltaic power, solar thermal, biomass power, and state-of-the-art fourth-generation nuclear power. It must also invest in distributed combined cooling, heating, and power (DCCHP) systems that use natural gas as fuel, and IGCC-CCS power generation, IGFC (hydrogasifier) power generation, low-cost CCS technology, and unconventional energy-use technologies of smart grids, energy storage, oil shale, and decarbonization and carbon storage.

2. Efficient, low-carbon and advanced energy end-use technologies

In order to improve energy efficiency, reduce carbon emissions, and encourage a reuse economy, it is necessary to develop advanced technologies for energy end-use sectors, including industry, transportation, and construction. These include advanced low-carbon steel production, electrolytic iron ore production, advanced dry-process preheat clinker production, cement kiln decarbonization, CCS technology, cascaded industrial waste energy and heat reutilization, new-fuel autos, pure electric vehicles, fuel-cell electric vehicles, efficient heat pump technology, solar thermal systems, and cogeneration systems using hydrogen fuel cells.

3. Alternative technologies for fuels and raw materials

In order to introduce alternative fuels, conserve raw materials, and utilize more biomass resources to reduce carbon emissions, it is necessary to develop such technologies as densified biofuels for household use, alcohol fuels from biomass, hydrocarbon fuels created from gasified biomass, and biodiesel created from biooil. Other important technologies include cement kilns that dispose of and recycle sludge and residential waste including plastics, and GHG reuse, such as the production of high-purity CO using CO_2 as gasifying agent, CO_2 flooding oil, and supercritical liquid CO_2 foaming.

8.2.4 Countermeasures and Strategic Studies

Key strategies and policy studies on climate change need to be emphasized to support low-carbon systems and sustainability in China.

1. Major national strategies and policies

It is necessary to develop and improve institutional, legal, policy, and evaluation frameworks on climate change, to determine appropriate strategies and policies on international trade to be adopted by China in the context of climate change, and to study how to develop technologies and techniques in support of the carbon emission trading market in China. It is also necessary to develop strategic measures and action plans on climate change adaptation, and to propose ambitious strategies for the development of cutting-edge science and technology on climate change and regional sustainable development.

2. International strategies and cooperation

It is necessary to study international political and economic actions in the context of climate change, in order to determine their impact on China's economy, trade, resources, energy, and ecological security. China must study the transfer of climate-friendly technologies and strategies for intellectual property protection and develop an integrated model that simultaneously addresses climate, economy, and society. It is necessary to analyze international and national long-term emission targets, reduction paths, mitigation and adaptation costs, and international climate conventions to determine how China can improve its international strategy on climate change, develop a polar protection strategy, and pursue international cooperative research on climate change.

3. Low-carbon and sustainable development

China must study green and low-carbon development and analyze potential impacts and social and economic costs and benefits of GHG emission reduction. The country needs to elaborate on emission-reduction strategies for industrialization, agricultural modernization, and urbanization, in order to create a plan for low-carbon development. It is necessary to study the impact of climate change on social development and consider regional, social, and economic impacts of climate change adaptation. China needs to launch climate change–related capacity-building and demonstration initiatives in key areas and regions and develop technologies and management practices for infrastructure and other high-profile projects. It is necessary to prepare national climate change policies on integrated technology demonstration zones or test beds for sustainable development.

References

IPCC. (2013). Climate change 2013: The physical science basis. In T. F. Stocker, D. Qin, G.-K. Plattner, M. Tignor, S. K. Allen, J. Boschung, A. Nauels, Y. Xia, V. Bex, & P. M. Midgley (Eds.), Contribution of Working Group I to the Fifth Assessment Report of the Intergovernmental Panel on Climate Change (1535 pp). Cambridge, UK: Cambridge University Press. doi:10.1017/CBO9781107415324

Olivier, J. G. J., Janssens-Maenhout, G., Peters, J. A. H. W., & Wilson, J. (2011). *Long-term trend in global* CO_2 *emissions*. 2011 Report, PBL/JRC, The Hague.

Printed in the United States
By Bookmasters